Gas Dehydration Field Manual

Gas Dehydration Field Manual

Maurice Stewart
Ken Arnold

ELSEVIER

AMSTERDAM • BOSTON • HEIDELBERG • LONDON • NEW YORK • OXFORD
PARIS • SAN DIEGO • SAN FRANCISCO • SINGAPORE • SYDNEY • TOKYO
Gulf Professional Publishing is an imprint of Elsevier

G | P
P |

Gulf Professional Publishing is an imprint of Elsevier
225 Wyman Street, Waltham, MA 02451, USA
The Boulevard, Langford Lane, Kidlington, Oxford, OX5 1GB, UK

Notices
Knowledge and best practice in this field are constantly changing. As new
research and experience broaden our understanding, changes in research
methods, professional practices, or medical treatment may become
necessary.

Practitioners and researchers must always rely on their own experience
and knowledge in evaluating and using any information, methods,
compounds, or experiments described herein. In using such information or
methods they should be mindful of their own safety and the safety of others,
including parties for whom they have a professional responsibility.

To the fullest extent of the law, neither the Publisher nor the authors,
contributors, or editors, assume any liability for any injury and/or damage
to persons or property as a matter of products liability, negligence or
otherwise, or from any use or operation of any methods, products,
instructions, or ideas contained in the material herein.

Library of Congress Cataloging-in-Publication Data
Application Submitted.

British Library Cataloguing-in-Publication Data
A catalogue record for this book is available from the British Library.

ISBN: 978-1-85617-980-5

For information on all Gulf Professional Publishing
publications visit our Web site at www.elsevierdirect.com

Printed and bound by CPI Group (UK) Ltd, Croydon, CR0 4YY

Contents

Part 1
Hydrate Prediction and Prevention

Contents

OBJECTIVES

To prepare natural gas for sale, its undesirable components (water, H_2S and CO_2) must be removed.

Most natural gas contains substantial amounts of water vapor due to the presence of connate water in the reservoir rock.

At reservoir pressure and temperature, gas is saturated with water vapor.

Removal of this water is necessary for sales specifications or cryogenic gas processing.

Primary concerns in surface facilities are determining the:

Water content of the gas

Conditions under which hydrates will form

Liquid water can form hydrates, which are ice-like solids, that can plug flow or decrease throughput.

Predicting the operating temperatures and pressures at which hydrate form and methods of hydrate prevention are discussed in this section.

DOI: 10.1016/B978-1-85617-980-5.00001-X

OVERVIEW

Dew Point

The dew point is the temperature and pressure at which the first drop of water vapor condenses into a liquid.

It is used as a means of measuring the water vapor content of natural gas.

As water vapor is removed from the gas stream, the dew point decreases.

Keeping the gas stream above the dew point will prevent hydrates from forming and prevent corrosion from occurring.

Dew Point Depression

Dew point depression is the difference between the original dew point and the dew point achieved after some of the water vapor is removed.

It is used to describe the amount of water needed to be removed from the natural gas to establish a specific water vapor content.

Why Dehydrate?

Dehydration refers to removing water vapor from a gas to lower the stream's dew point.

If water vapor is allowed to remain in the natural gas, it will:

Reduce the efficiency and capacity of a pipeline

Cause corrosion that will eat holes in the pipe or vessels through which the gas passes

Form hydrates or ice blocks in pipes, valves, or vessels

Dehydration is required to meet gas sales contracts (dependent upon ambient temperatures). Some examples include:

Southern U.S.A., Southeast Asia, southern Europe, West Africa, Australia   7 lb/MMSCFD

Northern U.S.A., Canada, northern Europe,

northern and central Asia   2–4 lb/MMSCFD

Cryogenic (turbo expander plants) 0.05 lb/MMSCFD

Solid bed adsorption units are used where very low dew points are required.

WATER CONTENT OF GAS

Introduction

Liquid water is removed by gas-liquid and liquid-liquid separation.

The capacity of a gas stream to hold water vapor is:

A function of the gas composition

Affected by the pressure and temperature of the gas

Reduced as the gas stream is compressed or cooled

When a gas has absorbed the limit of its water holding capacity for a specific pressure and temperature, it is said to be saturated or at its dew point.

Any additional water added at the saturation point will not vaporize, but will fall out as free liquid.

If the pressure is increased and/or the temperature decreased, the capacity of the gas to hold water will decrease, and some of the water vapor will condense and drop out.

Methods of determining the water content of gas include:

Partial pressure and partial fugacity relationships

Empirical plots of water content versus P and T

Corrections to the empirical plots above for the presence of contaminants such as hydrogen sulfide, carbon dioxide and nitrogen and Pressure Volume Temperature (PVT) equations of state.

Partial Pressure and Fugacity

Applying Raoult's law of partial pressures to water, we have

$$y_w = P_v x \tag{1-1}$$

Where:

y_w = Mol fraction of water in the vapor phase

P_v = Vapor pressure of water at system temperature

x_w = Mol fraction of water in the liquid water phase

= 1.0

Liquid mol fraction can be taken as unity because of the immiscibility of the liquid phases.

Therefore, for a known pressure and water vapor pressure the mol fraction water in the vapor phase can be determined from Equation 1-1.

Application of Equation 1-1

Valid only at low pressure where the ideal gas law is valid

Recommended for system pressures up to 60 psia (4 barg)

Empirical Plots

Empirical plots are based on lean, sweet natural gas.

The log of water content (w) is plotted versus P and T.

Plots approximate a straight line at a given pressure.

The water content shown is the maximum that gas can hold at the P and T shown.

It is fully saturated, that is, relative humidity is 100%.

The temperature is the water dew point temperature of the gas at the concentration and pressure shown.

Numerous correlations are available to determine the water content of a natural gas stream.

McKetta and Wehe correlation provides satisfactory results for most applications when used to determine the water content of a sweet natural gas stream that contains over 70% methane (Figure 1-1).

The accuracy is ± 5% (probably more accurate than the data the correlation is being applied toward).

As H_2S and CO_2 content increases the accuracy decreases. It is good practice to make corrections for these contaminants even though it may be small when concentrations and pressure are low.

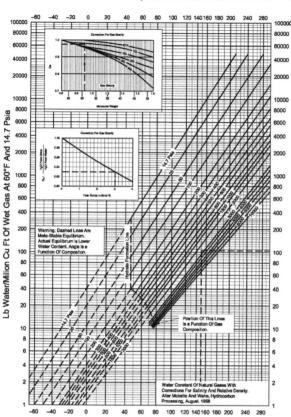

FIGURE 1-1 Water content of sweet, lean natural gas—McKetta-Wehe.

Comparison of the water content at various points in system serves many purposes:

Determine water loading for dehydration

Determine how much water has been condensed as liquid in the pipeline, which is:

Available to form hydrates

The root cause of many corrosion-erosion problems

Sour Gas Correlations

Weighted-Average Method

Correlation employs a *weighted average* to determine the water content of a gas stream containing sour gas.

In the approach, whereby the water content of the pure sour component is multiplied by its mol fraction in the mixture, the following equation can be used:

$$W = yW_{hc} + y_1W_1 + y_2W_2 \qquad (1\text{-}2)$$

Where:

W = Water content of gas

W_{hc} = Water content of hydrocarbon part obtained from McKetta-Wehe plot

W_1 = Water content of CO_2 obtained from appropriate empirical plot

W_2 = Water content of JH H_2S obtained from appropriate empirical plot

$y = 1 - (y_1y_2)$

y_1 = Mol fraction of CO_2

y_2 = Mol fraction of H_2S

Figures 1-2 and 1-3 show what is called the *effective water content.* Curves are based on the pure sour component data.

Sharma Correlation

The Sharma correlation utilizes Equation 1-2 and is based on the data obtained by Sharma.

Figures 1-4 and 1-5 were obtained by cross-plotting and smoothing Sharma's binary data for methane, CO_2, and H_2S.

SRK Sour Gas Correlation

The charts in Figure 1-6 were calculated from the SRK equation of state with the following assumptions:

The hydrocarbon portion of the gas was methane.

CO_2 had 75% of the water content of H_2S at the same conditions. One must multiply the percent CO_2 by 0.75 and add the result to the percent H_2S.

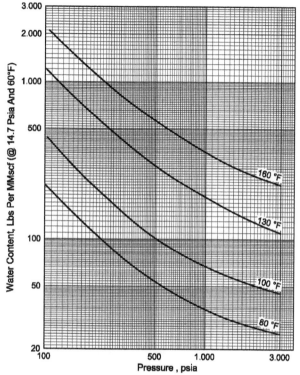

FIGURE 1-2 Effective water content of CO_2 in saturated natural gas mixtures.

The water content shown in API bbl/MMSCF can be converted as follows:

$$lbm/MMSCF = (350)(bbl/MMSCF)$$

The correlation is a "quick-look" way to estimate sour gas content.

Effect of Nitrogen and Heavy Ends

Nitrogen holds less water than methane.

Pressures up to 1000 psia (69 bara) water content of nitrogen is 5–10% less than methane.

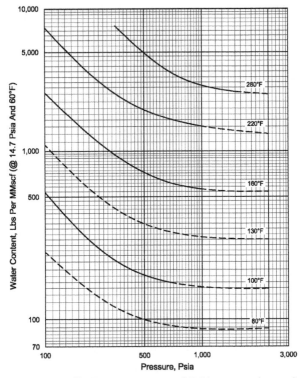

FIGURE 1-3 Effective water content of H_2S in saturated natural gas mixtures.

Deviation increases as pressure increases.

Including nitrogen as a hydrocarbon is practical and offers somewhat of a safety factor.

Presence of heavy ends tends to increase the water capacity of the gas.

Deviation is relatively small at normal system pressures.

The effects of nitrogen and heavy ends tend to cancel each other out in most systems.

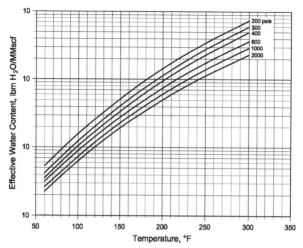

FIGURE 1-4 Water content of CO_2-Sharma.

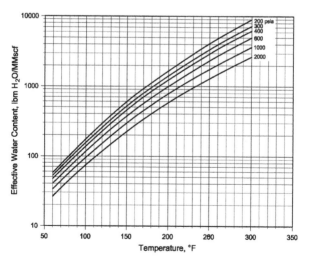

FIGURE 1-5 Water content of H_2S-Sharma.

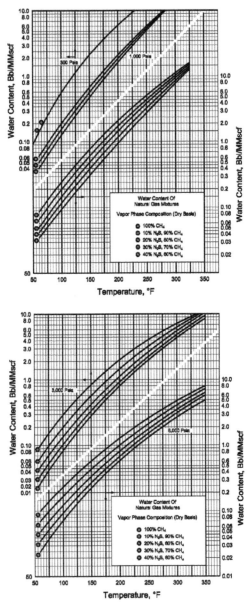

FIGURE 1-6 Sour gas correlation—SRK.

Example 1-1: Calculation of Water Content in a Sour Gas Stream

Determine: Calculate the saturated water content of a gas with the analysis shown below at 1100 psia and 120°F.

Composition	y_i
N_2	0.0046
CO_2	0.0030
H_2S	0.1438
C_1	0.8414
C_2	0.0059
C_3	0.0008
iC_4	0.0003
nC_4	0.0002
	1.0000

Solution:

1. From McKetta-Wehe plot read

   a. $W = 97$ lbm/MMSCF

2. From "effective water" content plots for CO_2 and H_2S, read

   a. $W_1 = 130$, $W_2 = 230$

   Substitute values into Equation 1-2:

   $$W = (0.8532)(97) + (0.003)(130) + (0.1438)(230)$$

   $$= 116 \text{ lbm/MMSCF}$$

3. From Sharma plots, $W_1 = 120$, $W_2 = 150$

   Substitute in Equation 1-2:

   $$W = (0.8532)(97) + (0.003)(120) + (0.1438)(150)$$
   $$= 105 \text{ lbm/MMSCF}$$

4. Effective percent of $H_2S = (\% CO_2)(0.75) + (\% H_2S)$

   $$= (0.3)(0.75) + (14.38)$$

   $$= 14.6\%$$

5. From SRK correlation plot, we must convert bbl/MMSCF to lbm/MMSCF

   $$W = (350)(\text{bbl/MMSCF})$$

   $$= (350)(0.31)$$

   $$= 109 \text{ lbm/MMSCF}$$

Note that the water content is greater using Equation 1-2 than from McKetta-Wehe plot.

A value of 116 is not likely but could happen.

Do not treat any one number as sacred. Look at the range when determining water content.

Applications

Correlations are used:

In dehydration calculations

To determine how much water, if any, will condense from the gas—involves considerations of disposal, corrosion/erosion and hydrate inhibition

Amount of Water Condensed

Need to make certain estimates are on the safe side of the possible range of values.

The additional capital expenditure will almost always be trivial.

There is a tendency to predict flowing temperatures lower than they turn out to be. The reason for this is the quality of the data used. Much of the data is obtained from a drill stem test, which is mediocre at best. Well-flowing temperature usually stabilizes upward after a few months in service.

The McKetta-Wehe plot is based on the log scale and thus a small change in temperature will result in a larger change in water content. For example, a 10% change in temperature results in a 33% increase in water content.

A common cause of poor dehydrator performance is under-predicting the water load.

GAS HYDRATES

What Are Gas Hydrates?

Gas hydrates are complex lattice structures composed of water molecules in a crystalline structure:

Resembles dirty ice but has voids into which gas molecules will fit

Most common compounds

Water, methane, and propane

Water, methane, and ethane

The physical appearance resembles a wet, slushy snow until they are trapped in a restriction and exposed to differential pressure, at which time they become very solid structures, similar to compacting snow into a snow ball.

Why Is Hydrate Control Necessary?

Gas hydrates accumulate at restrictions in flowlines, chokes, valves, and instrumentation and accumulates into the liquid collection section of vessels.

Gas hydrates plug and reduce line capacity, cause physical damage to chokes and instrumentation, and cause separation problems.

What Conditions Are Necessary to Promote Hydrate Formation?

Correct pressure and temperature and "free water" should be present, so that the gas is at or below its water dew point. If "free water" is not present, hydrates cannot form.

How Do We Prevent or Control Hydrates?

Add heat

Lower hydrate formation temperature with chemical inhibition

Dehydrate gas so water vapor will not condense into "free water"

Design process to melt hydrates

PREDICTION OF OPERATING TEMPERATURE AND PRESSURE

Wellhead Conditions

Temperature and pressure of a gas stream at the wellhead are important factors in determining whether hydrates will form when gas is expanded into the flowlines.

Temperatures at the wellhead increases as the flow rate increases and the pressure decreases.

Thus, wells that initially flowed under conditions causing hydrates to form in downstream equipment may decline out of the hydrate formation region as the reservoir depletes and the wellhead pressure drops.

Hydrate formation can sometimes be prevented if the flow rate from a well is maintained above some minimum rate.

This is an effective use of reservoir energy that would otherwise be lost in the pressure drop across a choke.

Flowline Conditions

The cooling of a gas in a flowline due to heat loss to the surroundings (ground, water, or air) can cause the gas temperature to drop below the hydrate formation temperature.

Records of flowline temperatures and pressures are needed to determine the best locations to effect pressure drops or install heaters.

Calculation of Temperature and Pressure at the Wellhead

Numerous computer programs are available that:

Calculate the temperature and pressure of a gas stream at the wellhead, and predict changes that will occur as the reservoir depletes

Calculation by hand is tedious and requires numerous iterations.

Calculation of Flowline Downstream Temperature

The conduction-convection equation can be used to calculate the downstream temperature of a flowline (T_d)

$$T_d = T_g + \frac{T_u - T_g}{e^x} \qquad (1\text{-}3)$$

Table 1-1 Heat Transfer Coefficients (U) for Various Bare Pipe Conditions (after Karge 1945)

Type Cover	Cover Condition	Depth of Cover (in.)	(Btu/hr/ft.³/°F)
	Dry	24	0.25–0.40
	Moist	24	0.50–0.60
	Soaked	24	1.10–1.30
	Dry	8	0.60–0.70
	Moist to wet	8	1.20–2.40
	Dry	24	0.20–0.40
	Moist	24	0.40–0.50
	Wet	24	0.60–0.90
	-	No soil cover	2–3
	Still	60 in. water plus	10
	River current	60 in. soil	2.0–2.5

Where:

T_d = flowline downstream temperature, °F

$$x = 24 \frac{(\pi DUl)}{(QC_p)}$$

D = flowline OD, ft.

U = heat transfer coefficient, Btu/hr/ft.²/°F
    = Table 1-1

L = flowline length, ft.

Q = gas flow rate, MSCFD

C_p = specific heat factor, Btu/MCF/°F = 26,800 normally used (Values in Table 1-2 multiplied by 1000 may yield more accurate results)

e = 2.718

T_u = upstream gas temperature, °F

(It could be the wellhead temperature (T_{WH}) if no choke or heater is used, or it could be the temperature downstream of a heater.)

T_g = ground temperature, °F
    = Table 1-3

TEMPERATURE DROP DETERMINATION

Overview

Choking (expansion of gas from high pressure to low pressure) is often required to control gas flow rates.

Table 1-2 Specific Heat Factor of 0.7 Specific Gravity Gas

Average Temperature in Flowline, °F	Flowline Pressure, psig											
	300	500	700	800	1000	1200	1500	1800	2100	2500	3000
120	29.1	30.3	31.0	31.6	32.5	33.3	34.8	36.2	37.2	28.8	40.6
100	28.7	29.9	30.8	31.4	32.4	33.4	35.1	36.7	38.0	39.7	41.6
80	28.2	29.5	30.5	31.3	32.4	33.5	35.4	37.2	38.7	40.5	42.5
60	27.5	29.2	30.3	31.1	-	-	-	-	-	-	-

After National Tank Company (1958)

Table 1-3 Average Ground Temperatures (T_g), °F

Cover, in.	T_g, °F
36	53 to 58
18	25 to 45 (Northern Europe, Canada, Alaska)
	45 to 48 (Northern U.S.A., China, Russia)
	48 to 53 (Southern U.S.A., Southeast Asia, West Africa, South America)

Chokes and control valves are commonly used.

Pressure drop across the restriction causes a decrease in gas temperature.

If the gas is saturated with water and the final temperature of the gas is below the hydrate formation temperature, then hydrates will form.

Pressure drop across a choke is a constant enthalpy process.

For a multicomponent stream, one must perform flash calculations which balance enthalpy before and after the choke. This is better suited for a computer.

Temperature Drop Correlation (Figure 1-7)

Used when gas composition is unknown

Used for a "first approximation"

Yields reliable results but affected by liquids

Requires correcting for hydrocarbon liquids

Accuracy is ±5%.

Example 1-2: Determine the Temperature Drop across a Choke

Given: A well with a flowing tubing pressure of 4000 psi and 20 bbl of hydrocarbon condensate and a downstream back pressure of 1000 psi.

Solution: Initial pressure = 4000 psi

Final pressure = 1000 psi

ΔP = 3000 psi

From Figure 1-7 correlation; intersect initial pressure = 4000 and ΔP-3000 read ΔT = 80°F.

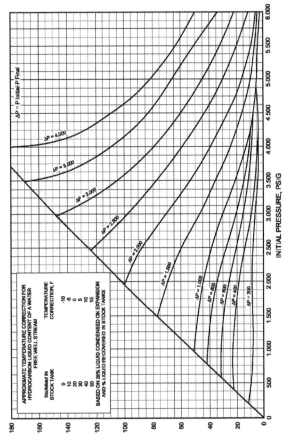

FIGURE 1-7 Temperature drop accompanying a given pressure drop for a natural gas stream.

HYDRATE PREDICTION CORRELATIONS

Overview

All correlations are based on a system that contains only gas and water in a static test cell that was rocked only to provide good equilibrium.

Data shown are the hydrate melting conditions, not the formation point

Yield acceptable results

Correlations are used to predict hydrate formation temperature.

Vapor-Solid Equilibrium Constants

Yield reliable results up to 1000 psia

Used with composition of the stream is known

Pressure-Temperature Curves

Results are not as accurate as vapor-solid equilibrium constants.

Used when composition of the stream is not known.

Used for "first approximations" or "quick look."

Equations of State Calculations

Computer solutions developed to predict hydrate formation conditions.

Vapor-Solid Equilibrium Constants

This procedure is used to determine hydrate formation temperature when the stream composition is known.

1. Assume hydrate formation temperature

2. Determine the equilibrium constant, K, for each component where

$$K_i = \frac{Y_i}{X_i} \qquad (1\text{-}4)$$

Where:

$Y_i =$ Mol fraction of each component in the gas on a water free basis

$X_i =$ Mol fraction of each component in the solid on a water free basis

3. Calculate the ratio, Y_i/K_i, for each component

4. Sum the values of Y_i/K_i

5. Repeat steps 1–4 for additional temperatures
    until $\sum Y_i/K_i = 1$

Figures 1-8 through 1-12 are graphs giving vapor-solid
equilibrium constants, K, at various pressure and
temperatures.

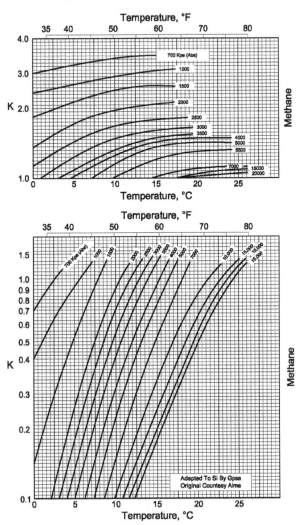

FIGURE 1-8 Vapor-solid "K" values for methane and ethane.

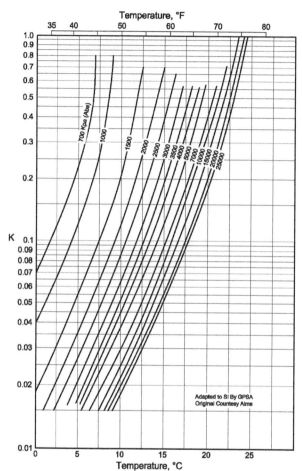

FIGURE 1-9 Vapor-solid "K" values for propane.

Gas streams containing more than 30% H_2S behave as if they contained pure H_2S.

Components heavier than butane have $K_i = $ infinity, since their molecules are too large to fit into the cavities of the lattice structure.

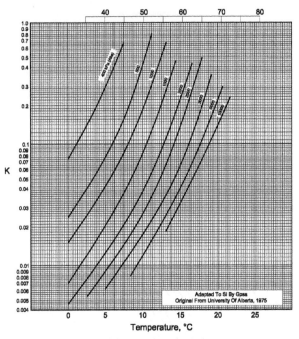

FIGURE 1-10 Vapor-solid "K" values for isobutane.

Example 1-3: Determination of Hydrate Formation Temperature Using Vapor-Solid Constants

Given: A flow stream with a flowing pressure of 400 psia and the following composition.

Determine: The hydrate formation temperature.

Component	Mole Fraction of Gas
Nitrogen	0.0144
Carbon Dioxide	0.0403
Hydrogen Sulfide	0.000019
Methane	0.8555
Ethane	0.0574
Propane	0.0179
Isobutane	0.0041
n-Butane	0.0041
Pentane +	0.0063
	1.00000

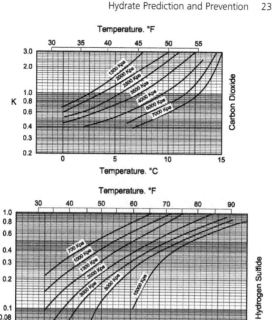

FIGURE 1-11 Vapor-solid "K" values for CO_2 and H_2S.

Solution:

Calculation of Temperature for Hydrate Formation at 400 psia

Component	Mole Fraction in Gas	At 70°F		At 80°F	
		K_i	Y_i/K_i	K_i	Y_i/K_i
Nitrogen	0.0144	Infinity	0.00	Infinity	0.00
Carbon Dioxide	0.0403	Infinity	0.00	Infinity	0.00
Hydrogen Sulfide	0.000019	0.3	0.00	0.5	0.00
Methane	0.8555	0.095	0.90	1.05	0.81
Ethane	0.0574	0.72	0.08	1.22	0.05
Propane	0.0179	0.25	0.07	Infinity	0.00
Isobutane	0.0041	0.15	0.03	0.06	0.01
n-Butane	0.0041	0.72	0.00	1.22	0.00
Pentane +	0.0063	Infinity	0.00	Infinity	0.00
Total	**1.0000**		**1.08**		**0.87**

NOTES:
Interpolating linearly, V/K = 1.0 at 74°F
Therefore, hydrates will form at 75°F

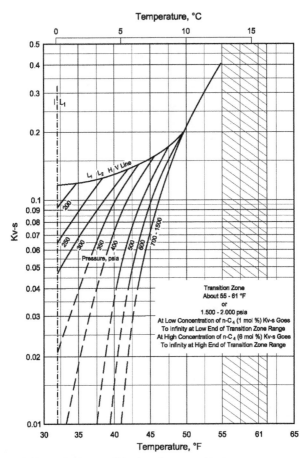

FIGURE 1-12 Vapor-solid values for normal-butane.

Pressure-Temperature Curves (Figure 1-13)

Pressure-temperature curves are used when gas composition is not known or for "first approximation."

Graphs have been developed to approximate hydrate formation temperature as a function of

Gas gravity

Pressure

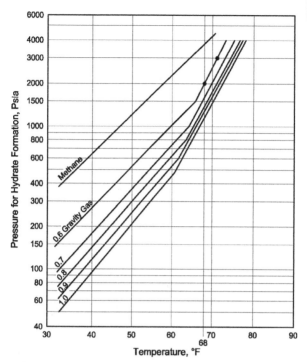

FIGURE 1-13 Pressure-temperature curves for predicting hydrate formation temperature.

Example 1-4: Determine the Hydrate Formation Temperature Using Pressure-Temperature Correlations

Given: A 0.6 specific gravity gas operating at 2000 psia.

Solution: From pressure-temperature curve intersect 2000 psia and 0.6 specific gravity curve and read 68°F.

HYDRATE PREVENTION

Overview

Hydrate prevention is used to prevent hydrates from forming.

Operating conditions must remain out of the hydrate-formation zone

Hydrate point must be maintained below the operating conditions of the system

Two common methods of hydrate-formation prevention are:

Temperature control

Chemical injection

Adding Heat

Adding heat is effective because hydrates normally do not occur above 70°F.

It offers a simple and economical solution for land and offshore facilities (if waste heat is available).

Flow stream is preheated, either through an indirect line heater or heat exchanger, before passing through a choke. Flow stream is then reheated to maintain the temperature above the hydrate formation temperature.

A major drawback in offshore installations is that it is almost impossible to maintain flowline temperatures significantly above the water temperature if the flowlines extend more than a few hundred feet under water.

Thus, either the "free water" must be separated while still at temperature or an alternate method selected.

Temperature Control

Indirect Heaters
Overview

An indirect heater is used to heat gas to maintain temperatures above that of the hydrate formation.

It consists of an atmospheric vessel containing a fire tube (usually fired by gas, steam, or heating oil) and a coil (designed to withstand SITP) that is heated by the intermediate fluid (usually water) and the fluid is heated.

The fire tube and coil are immersed in a heat transfer fluid (normally water), and heat is transferred to the fluid in the coil.

Wellhead Heater Description *(Figures 1-14 and 1-15)*

Figure 1-14 shows a typical heater installation at the wellhead.

Beginning at the wellhead, the following items are normally included:

Safety shut-down "wing" valve

Pneumatically actuated valve that is connected directly to the Christmas tree

PSL pilot, will shut-in the well whenever the flowline pressure upstream of the heater falls below a certain set pressure, indicative of a flowline rupture

PSHL pilot, senses flowline pressure downstream of the heater choke and will shut-in the well on either abnormally high or low pressure

High-Pressure Flowline

A line, normally at least 150 feet long, designed to withstand full wellhead Shut-In Tubing Pressure (SITP).

Expansion Loop

A loop designed to absorb flowline length changes caused by changes in temperature between flowing and shut-in conditions.

Long-Nose Heater Choke *(Figure 1-15)*

A long body choke installed in the indirect heater to position the choke orifice within the indirect heater bath.

Since the walls of the choke orifice are heated by the water bath, hydrates will not form in the orifice and cause plugging.

Heater Bypass Valve

A valve designed to withstand full wellhead shut-in pressure and bypass gas around the heater after the wellhead pressure has been drawn down to near sales line pressure.

Use of this valve prevents needless wear and erosion on the heater coils and allows the pressure drop from wellhead to sales line to be minimized.

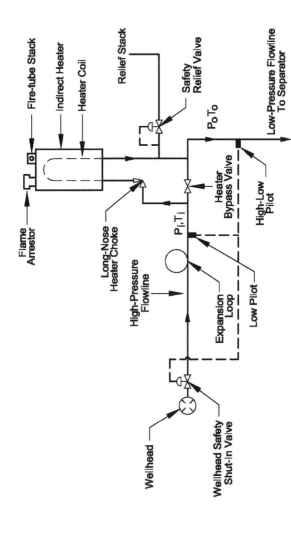

FIGURE 1-14 Typical wellhead indirect heater schematic.

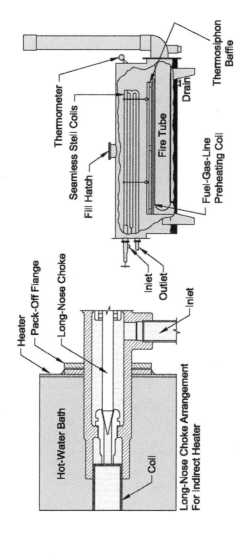

FIGURE 1-15 Indirect heater details.

Heater Coil

A multiple-pass steel coil designed to withstand full wellhead SITP.

Because of high corrosion and erosion rates, the return bends are "safety-drilled."

Holes will begin to leak when corrosion and erosion have reduced the wall thickness by half, warning that the bends should be replaced.

Pressure Relief Valve

Provides over-pressure protection for the low-pressure flowline.

Heater Flame Arrestor

A device that provides fire protection by preventing the heater fire from flashing back through the air intake and igniting surrounding material.

Flowline Heaters

Flowline heaters differ from wellhead heaters in purpose only.

The purpose of a wellhead heater is to heat the flow stream at or near the wellhead where choking or pressure reduction occurs.

The purpose of a flowline heater is to provide additional heat if required.

The design is the same as an indirect heater except that the choke, shut-in, and relief equipment are seldom used.

A bypass should be installed in either case so that the heater can be taken out of service.

System Optimization

System operation has to be optimized before heaters can be effectively designed and located.

Heat requirements that appear to be large can often be reduced to minimal values or even eliminated by revising the mode of operation. For example:

Fields having multiple producing wells can be combined to use higher flowing temperatures thus minimizing heater requirements.

If reducing the gas stream pressure is necessary, it is generally more efficient to do so at a central point where the necessary heater fuel gas can be obtained from separators or scrubbers.

Requires flowline wall thickness to be increased so as to withstand wellhead SITP.

An alternative is to install wellhead shut-down valves and flowline PSHs.

Heater Sizing

To adequately describe the size of a heater, heat transfer duty and coil sizes must be specified.

To determine the heat duty required, we must know:

Amount of gas, water, and oil or condensate expected

Pressures and temperatures of the heater inlet and outlet

The heater outlet temperature depends on the temperature at which hydrates form.

The coil size depends on:

Volume of fluid flowing through the coil

Required heat-transfer duty

Special operating conditions such as start-up of a shut-in well should be considered.

Downhole Regulators

Downhole regulators are feasible for high capacity gas wells at locations where certain risks to other downhole equipment are acceptable.

The theory behind the use of a downhole regulator is that the pressure drop from flowing pressure to near-sales line pressure is taken downhole where the formation temperature is sufficient to prevent hydrate formation. The tubing string above the regulator then acts as a subsurface heater.

Calculations involved in downhole regulator design are rather involved. They depend on characteristics such as:

Wellbore configuration

Flowing downhole pressures and temperature

Well depth

Although shortcut procedures are available to estimate the feasibility of downhole regulators, tool company representatives can provide detailed design information.

Chemical Injection

Overview

Hydrate inhibitors are used to lower the hydrate formation temperature of the gas.

Methanol and ethylene glycol are the most commonly used inhibitors.

Recovery and regeneration steps are used in all continuous glycol injection projects and in several large-capacity methanol injection units.

Injection of hydrate inhibitors should be considered for the following applications:

Pipeline systems in which hydrate trouble is of short duration

Gas pipelines that operate at a few degrees below the hydrate formation temperature

Gas-gathering systems in pressure-declining fields

Gas lines in which hydrates form as localized points

Methanol and the lower molecular weight glycols have the most desirable characteristics for use at hydrate inhibitors.

Table 1-4 lists some physical properties of methanol and the lower molecular weight glycols.

When hydrate inhibitors are injected in gas flowlines or gathering systems, installation of a free-water knockout (FWKO) at the wellhead proves to be economical in nearly every case.

Removing the free water from the gas steam reduces the amount of inhibitor required.

Methanol Injection Considerations

Methanol is well-suited for use as a hydrate inhibitor because it is:

Noncorrosive

Nonreactive chemically with any constituent of the gas

Soluble in all proportions in water

Volatile under pipeline conditions

Reasonable in cost

Of a vapor pressure greater than that of water

Table 1-4 Physical Properties of Chemical Inhibitors

Property	Methanol	Ethylene Glycol	Diethylene Glycol	Thiethylene Glycol	Tetraethylene Glycol
Molecular weight	32.04	62.10	106.10	150.20	194.23
Boiling point at 760 mm Hg, °F	148.10	387.10	427.60	532.90	597.2
Vapor pressure at 77°F, mm Hg	94	0.12	<0.01	<0.01	<0.01
Specific gravity at 77°F	0.7868	1.110	1.113	1.119	1.120
Specific gravity at 140°F	–	1.085	1.088	1.092	1.092
Pounds per gallon at 77°F	6.55	9.26	9.29	9.34	9.34
Freezing point, °F	−144	8	17	19	22
Pour point, °F	–	<−75	−65	−73	−42
Absolute viscosity in centipoises at 77°F	0.55	16.5	28.2	37.3	39.9
Absolute viscosity in centipoises at 140°F	0.36	5.1	7.6	9.6	10.2
Surface tension at 77°F, dynes/cm	22	47	44	45	45
Specific heat at 7°F, Btu/lb/°F	0.27	0.58	0.55	0.53	0.52
Flash pont, °F	0	240	280	320	365
Fire point, °F	0	245	290	330	375
Decomposition temperature, °F	0	329	328	404	460
Heat of vaporization at 14.65 psi, Btu/lb	473	364	232	179	–

Methanol Injection System Description
(Figure 1-16)

> Methanol is injected by means of a gas-driven pump (3 in Figure 1-16) into the flowline upstream of the choke or pressure control valve (2).
>
> A temperature controller (5) measures the temperature of the gas in the low-pressure flowline (7) and adjusts the methanol rate accordingly.
>
> The methanol injection rate is controlled by the amount of power gas allowed to flow through the power gas control valve (4) to drive the pump.

Glycol Injection Considerations

> Glycol has a relatively low vapor pressure and thus does not evaporate into the vapor phase as readily as methanol.
>
> The solubility of glycol in liquid hydrocarbons is relatively low.
>
> For the above reasons, glycol can be more economically recovered, thus reducing the operating expenses below those of methanol systems.

Glycol Injection and Recovery System Description

> The injection part of the system (items 1 to 5 in Figure 1.17) is similar to the methanol injection system.
>
> Additional equipment in the glycol system is for recovering and reclaiming the glycol.
>
> A three-phase separator (6) separates the water and glycol from the hydrocarbon phases.
>
> The water–glycol solution in the separator is sent to the reboiler (7) while gas is delivered to the sales line, and the hydrocarbon condensate is dumped to the condensate storage tanks.
>
> In the reboiler, excess water is boiled away from the glycol.
>
> The glycol reconcentrated in the reboiler is then available again for injection into the gas stream.
>
> Separation of the glycol water phase from the hydrocarbon-liquid field requires a temperature above 70°F and a residence time of 10 to 15 minutes.

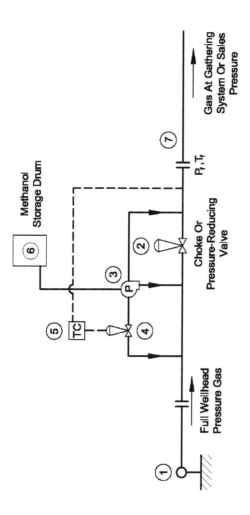

FIGURE 1-16 Typical methanol injection system.

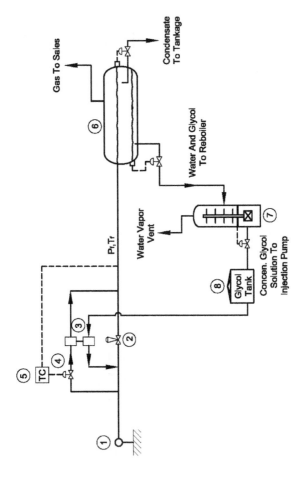

FIGURE 1-17 Typical glycol injection and recovery system.

Nozzle Design (Figure 1-18)

Due to the vapor pressure of glycol, a fine, well-distributed mist is required to obtain adequate mixing with the gas to ensure optimum results, thus spray nozzles are normally used.

Nozzle selection is a major consideration in the design of cold separation facilities or plants using glycol injection.

Glycol injection normally takes place just upstream of a heat exchanger or chiller where gas is being chilled.

Proper nozzle selection will ensure that the glycol spray covers the tube sheet.

100 to 150 psi differential pressure at the nozzle is sufficient to atomize the glycol.

Process stream velocities should be at least 12 fps.

Glycol Selection

The three glycols normally used to prevent the formation of hydrates are:

Ethylene glycol (EG)

Diethylene glycol (DEG)

Triethylene glycol (TEG)

Selection of a glycol depends on the composition of the hydrocarbon flow stream and the advice of the glycol supplier.

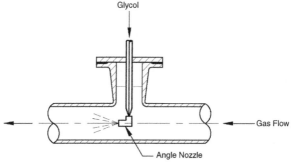

FIGURE 1-18 Schematic of a spray nozzle used in glycol injection.

Dow Chemical Company Guidelines

If glycol is to be injected into a natural gas transmission line where glycol recovery is of less importance than hydrate protection, ethylene glycol is the best choice because it produces the greatest hydrate depression and has the highest vapor pressure of any of the glycols.

If glycol is to be injected into a unit where it will contact hydrocarbon liquids, ethylene glycol is preferred because it has the lowest solubility in high molecular weight hydrocarbons.

If vaporization losses are severe, either diethylene or triethylene glycol are the best choice because both have a lower vapor pressure. Sometimes diethylene glycol is used if there is a combined loss of both gas vaporization and liquid solubility.

The freezing point of the glycol solution must be lower than the lowest temperature expected in the system. In inhibitor service, glycol concentrations are usually maintained at 70 to 75 weight percent because freezing of the glycol is not a problem at this concentration.

Reboiler temperature is dependent on the type of glycol and its concentration.

> Temperature should be maintained at a level equal to the boiling point of the desired solution.

Boiling points for the three glycol types are plotted in Figures 1-19, 1-20, and 1-21.

For example, from Figure 1-19 the reboiler temperature should be set at 240°F in order to produce a 70 weight percent ethylene glycol solution at atmospheric pressure (760 mm). Thermal degradation can occur if the boiling point of the pure glycol is exceeded; it should therefore be avoided.

Glycol losses for the two-phase gas condensate systems are normally estimated at 1 to 2 gallons per 100 barrels of hydrocarbon liquid produced.

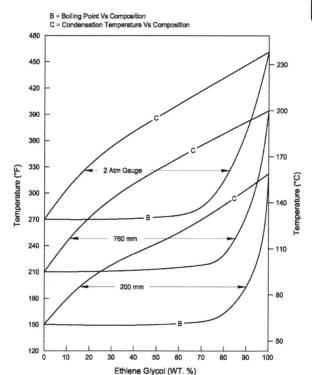

B = Boiling Point Vs Composition
C = Condensation Temperature Vs Composition

FIGURE 1-19 Boiling points and condensation temperatures of aqueous ethylene glycol solutions at various pressures.

Vaporization into the gas stream and solution into the hydrocarbon liquid usually cause only a small portion of the total loss.

The most significant causes of glycol losses are leakage and carryover with the hydrocarbon liquid.

Loses also occur from vaporization and carry over in the reboiler.

Injection Requirement Categories
Low Pressure–High Volume

Pressures up to 2000 psi and volumes measured in hundreds to thousands of BPD.

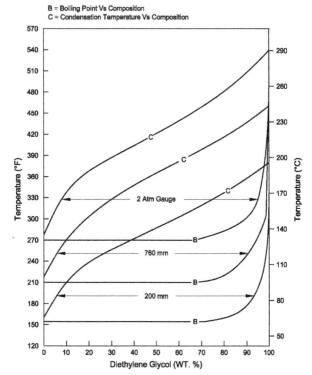

B = Boiling Point Vs Composition
C = Condensation Temperature Vs Composition

FIGURE 1-20 Boiling points and condensation temperatures of aqueous diethylene glycol solutions at various pressures.

High Pressure–Low Volume

Pressures up to 15,000 psi and volumes measured in quarts or a few gph.

High Pressure–High Volume

Pressures exceeding 5000 psi and volumes measured in several gpm or barrels per minute. This is the most difficult to deal with.

Intermittent control when problems are encountered as "local" heat can be applied at the surface to remove hydrate plugs that may occur.

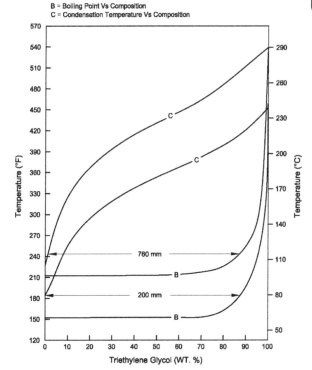

B = Boiling Point Vs Composition
C = Condensation Temperature Vs Composition

FIGURE 1-21 Boiling points and condensation temperatures of aqueous triethylene glycol solutions at various pressures.

Single versus Two-Step Injection Considerations

Single-Step Injection

All chemicals are injected through a downhole tubing mandrel that handles both wellhead and flowline conditions.

Two-Step Injection

Utilizes a second injection point just downstream of the wellhead to handle additional water that condenses from the gas phase as the flow stream cools to the surrounding temperature.

Chemical Injection System

Overview

The three parts of an injection system are:

Pump

Meter

Control system

Single-Point Chemical Injection

A single pump, meter, and control system service one injection point.

Disadvantages

Limited turn-down capability and increased life-cycle cost

Weight and space increase as injection points increase

Multi-Point Chemical Injection

A shared pump and multiple meter and control devices servicing multiple injection points.

Advantages

Increased turn-down capacity and inferred flow monitoring

Self-compensated with closed loop control

Per well capital investment decreases as the number of wells increases

Injection points are easily added

Lower weight and space requirements for higher quantity well applications

Disadvantages

Instrumentation intensive

Multiple control loops required

Requires variable speed for fixed crank pumps

Experiences high-pressure drops from header to recycle line

Metering Pump Considerations

Features pump-meter-control functions

Vertical or horizontal

Variable crank

Robust design

Modular construction

Pumphead interchangeability

Highly precise and repeatability

Covered in API 675

Diaphragm Pumps
Advantages

Hermetically sealed, no contamination to atmosphere

Long-life diaphragms typically greater than 2 years continuous duty (20,000 hours)

Long-life of hydraulic plunger seals, typically greater than 2 years continuous duty (20,000 hours)

Internal hydraulic relief

Maximum safeguard to environment and personnel safety automated diaphragm failure mechanisms

Disadvantages

Higher purchase price (pay back in less down time)

More complex maintenance required

Plunger Pumps
Advantages

Lower purchase price

Less complicated maintenance (easier to understand)

Disadvantages

Plunger packing service life typically less than 2000 hours

Friction between plunger and packing

Comparison of Hydrate Prevention Methods

Overview
The four methods (indirect heaters, methanol injection, glycol injection, and downhole regulators) discussed above are proven safe and reliable.

Evaluation should consider:

Development of CAPEX and OPEX (including chemicals and fuel)

Space needs (especially in offshore operations) and operating hazards.

Heaters

Capital costs and the fuel expense of heaters are relatively large, and it is difficult to maintain a clean, reliable fuel supply to remote heater locations.

Indirect heater requires a large amount of space.

Fire boxes with proper flame arrestors have minimized the hazards from fired equipment, but they should be bought with strict attention paid to detailed design.

Chemical Injection

Advantages and disadvantages of methanol injection and glycol injection are listed in Table 1-5.

The use of methanol requires only a free-water separator and a suitable means for injection and atomizer, whereas the use of glycol requires a free-water separator plus a gas–liquid separator and a glycol reconcentration unit at the point of recovery downstream.

Downhole Regulators

No routine service is required on downhole regulators, but a wireline service company must be used each time the pressure drop has to be changed and when the regulator is removed.

A well with a downhole regulator may require injection of methanol or glycol when it is brought back online after a shut-in until the flow and temperature stabilize.

After a well declines to less than allowable production, the downhole regulator will have to be removed, and another form of hydrate prevention may prove necessary.

Table 1-5 Methanol and Glycol Injection Comparisons

Inhibitor	Advantages	Disadvantages
Methanol	Relatively low initial cost Minimal equipment Simple system with little gas consumption	High operating cost Hauling to site necessary
Glycol	Usually lower operating cost than methanol when both systems recover chemical Simple system with little gas consumption	High initial cost Hauling to site necessary Large loss if line breaks Possibility of glycol concentration

Downhole regulators do not present special safety hazards, but because work with regulators involves working in the well, losing the well is always a danger.

Summary of Hydrate Prevention Methods

The methanol injection system is often used for temporary hydrate prevention service in small installations.

Larger installations are favored for indirect heaters or glycol injection systems.

Downhole regulators are most useful in large high-pressure reservoirs in which excess pressure is available and the reservoir pressure is not expected to decline rapidly.

Table 1-6 contains a summary comparison of the above methods.

HYDRATE INHIBITION

Hammerschmidt Equation

The Hammerschmidt equation is used to determine the amount of inhibitor required in the water phase to lower the hydrate temperature. It is expressed as:

$$\Delta T = \frac{KW}{100(MW) - (MW)(W)} \tag{1-5}$$

Where:

ΔT = Depression of hydrate formation temperature, °F

MW = Molecular weight of inhibitor

K = Constant, from table below

W = Weight percent of inhibitor in final water

Inhibitor	Constants	
	MW	K
Methanol	32.04	2335
Ethanol	46.07	2335
Isopropanol	60.10	2335
Ethylene Glycol	62.07	2200
Propylene Glycol	76.10	3540
Diethylene Glycol	106.10	4370

Table 1-6 Comparison of Hydrate Prevention Methods

Technique	Investment	Fuel	Operating Maintenance	Chemicals	Plot Area	Hazards	Downtime
Downhole regulators	Very low	None	Low	None	None	High	Low
Wellhead heaters	Very high	Very high	Low	Very low	Very high	High	Low
Methanol injection	Very low	None	Low	Very high	Very low	Medium	Low
Glycol injection	High	Medium	Low	High	Very high	High	Low

Determination of Total Inhibitor Required

$$\begin{pmatrix} Total \\ inhibitor \\ required \end{pmatrix} = \begin{pmatrix} Inhibitor \\ required \\ free\ water \end{pmatrix} + \begin{pmatrix} Inhibitor \\ lost\ to \\ vapor\ phase \end{pmatrix}$$

$$+ \begin{pmatrix} Inhibitor \\ soluble \\ condensate \end{pmatrix} \qquad (1\text{-}6)$$

Where:

Inhibitor lost to vapor phase is determined from Figure 1-24. Methanol lost to the vapor phase

Inhibitor soluble in the condensate is approximately 0.5%

Procedure for Determining Inhibitor Requirements

Best illustrated by an example. Refer to Figure 1-22.

Example 1-5: Determining the Amount of Methanol Required in a Wet Gas Stream

Given: FWHT = 100°F for the subsea well

Determine: Calculate the total methanol required to prevent hydrates from forming. A conservative approach is to assume that the gas is saturated at wellhead conditions.

Solution:

1. The amount of water that will be condensed is determined from McKetta-Wehe (Figure 1-23), assuming the gas is saturated at reservoir and wellhead conditions.

Water Content	= 32.0 lb/MMSCF
(@ 3,000 psia & 100°F)	(@ wellhead)
Water Content	= -11.5 lb/MMSCF
(@ 2,000 psia & 60°F)	(@ platform)
Water Condensed	= 20.5 lb/MMSCF
Produced Water	= +1083 lb/MMSCF
Total	= 1103.6 lb/MMSCF

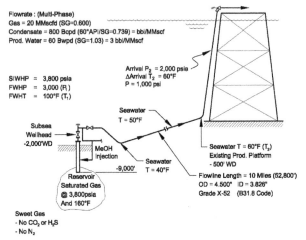

Subsea Well Tie-back To An Existing Shallow Water Production Platform

Flowrate : (Multi-Phase)
Gas = 20 MMscfd (SG=0.600)
Condensate = 800 Bcpd (60°API/SG=0.739) = bbl/MMscf
Prod. Water = 60 Bwpd (SG=1.03) = 3 bbl/MMscf

Arrival P_2 = 2,000 psia
ΔArrival T_2 = 60°F
P = 1,000 psi

SIWHP = 3,800 psia
FWHP = 3,000 (P_1)
FWHT = 100°F (T_1)

Seawater
T = 50°F

Subsea
Wellhead
-2,000'WD

MeOH
Injection

Seawater
-9,000'   T = 40°F

Seawater T = 60°F (T_2)
Existing Prod. Platform
- 500' WD

Flowline Length = 10 Miles (52,800')
OD = 4.500"   ID = 3.826"
Grade X-52   (B31.8 Code)

Reservoir
Saturated Gas
@ 3,800psia
And 160°F

Sweet Gas
- No CO_2 or H_2S
- No N_2

FIGURE 1-22 Subsea methanol injection example.

2. From pressure-temperature curve, the hydrate
    formation temperature is 68°F (refer to
    Figure 1-13)

    The required dewpoint depression then is
    68°F–60°F = 8°F

3. The concentration of methanol required in the
    liquid water phase from Equation 1-5 is

    $$8°F = \frac{2335W}{100(32.042) - (32.042)W}$$

    Rearranging and solving for W = 9.892% = 0.
    09892

4. Therefore, the estimated methanol required in the
    liquid water phase is:

    $$= \frac{0.09892}{1 - 0.09892}(1103.65 \text{ lb/MMSCF})$$

    $$= 121.15 \text{ lb/MMSCF}$$

5. From Figure 1-24, the methanol that will flash into
    the vapor phase at 2000 psia and 60°F is:

    $$= \frac{(x) \text{ lbs.methanol/MMSCF(@14.7 psia \& 60°F)}}{WT \text{ \& methanol in water phase}}$$

    $$= 1.52$$

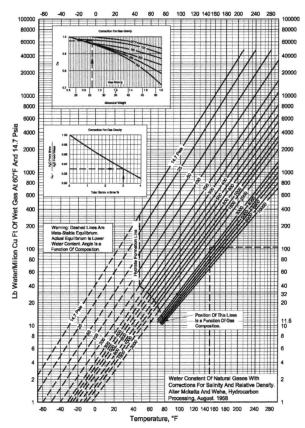

FIGURE 1-23 Water content of sweet, lean natural gas—McKetta-Wehe example.

6. Therefore, the methanol in the vapor phase (x) is:

$$(1.52)(9.892\%) = 14.94\,lb/MMSCF$$

7. A barrel of our condensate weighs:

$$(0.739)(5.6146\,cf/bbl)(62.41\,lb/cf) = 258.9\,lb/bbl$$

8. Therefore, the approximate amount of methanol soluble in the condensate or liquid hydrocarbon phase (assuming a 0.5% solubility by weight) is:

$$= \frac{0.005}{1 - 0.005}(258.9\,lb/bbl)(40\,bbl/MMSCF)$$

$$= 52.04\,lb/MMSCF$$

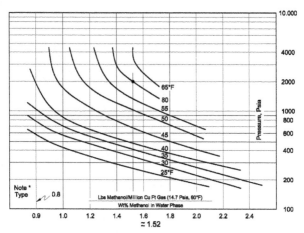

FIGURE 1-24 Ratio of methanol vapor composition to methanol liquid composition-example.

9. Thus, the total amount of methanol required is:

Liquid water phase   = 121.15 lb/MMSCF

Vapor phase   = 14.94 lb/MMSCF

Soluble in condensate = <u>52.04</u> lb/MMSCF

Total   = 188.13 lb/MMSCF

(188.13 lb/MMSCF)
(20 MMSCFD)   = 3762.6 lb/day

Note that for gas-condensate wells producing a reasonable or high amount of condensate, the amount of methanol soluble in the condensate is crucial to determining the amount needed.

Approximately 188 lb of methanol must be added so that approximately 121 lb will be dissolved into the water phase. Since the specific gravity of methanol is 0.791 (at 68°F), this is equivalent to:

$$\frac{188.13 \text{ lb/MMSCF}}{(0.791)(8.3453 \text{ lb/gal})} = 28.5 \text{ gal/MMSCF}$$

$$\frac{(28.5 \text{ gal/MMSCF})}{42 \text{ gal/bbl}} = 0.679 \text{ bbl/MMSCF}$$

(0.679 bbl/MMSCF)(20 MMSCF) = 13.57 bbl/day

Hydrate Inhibition-Hammerschmidt's Equation
Minimum Inhibitor Requirements-Methanol
Works Rev.3: GPSA (Fig. 20-33, p. 20-18)

Note the sensitivity of the overall MeOH
volume based on amount of condensate
and the solubility of the MeOH into the
condensate liquid phase!

Input Values:

	CASE I.D.:	Well #1	Well #1	Well #1	Well #1	Well #1
Qg = Nominal Gas Flow Rate (MMscfd)		20	20	20	20	20
Ql = Condensate Flow Rate (BBL/Day)		800	800	800	800	800
Qw = Free Water Flow Rate (BBL/Day)		60	60	60	60	60
SGg = Gas Specific Gravity		0.600	0.600	0.600	0.600	0.600
SGc = Condensate Specific Gravity		0.739	0.739	0.739	0.739	0.739
SGw = Free Water Specific Gravity		1.030	1.030	1.030	1.030	1.030
SGm = Methanol Specific Gravity		0.791	0.791	0.791	0.791	0.791
Wc% = MeOH Solubility In Condensate (wt%)		3.00	1.50	1.00	0.80	0.53
W1 = Sat. Water Content Upstream Of Choke (Lb/MMscf)		32.0	32.0	32.0	32.0	32.0
W2 = Sat. Water Content Downstream Of Choke (Lb/MMscf)		11.5	11.5	11.5	11.5	11.5
P2 = Pressure Downstream Of Choke (PSIG)		2,000	2,000	2,000	2,000	2,000
T2 = Temperature Downsteam Of Choke (°F)		60.0	60.0	60.0	60.0	60.0
dT = Desired Temperature Margin (°F) Above Hydrate Formation Temperature (ie, Safety Factor)		0	0	0	0	0

Intermediate Values:

Wf = Free Water (Lb/day)	21,653	21,653	21,653	21,653	21,653
Wc = Water Condensed (Lb/Day)	410	410	410	410	410
Wo = Total Water Flow Rate	22,063	22,063	22,063	22,063	22,063
R = Methanol Vapor/Liquid Ratio (Lbs MeOH/MMscf/wt%)	1.508	1.508	1.508	1.508	1.508
d = Req'd. Water Dewpoint Depression (°F)	8.35	8.35	8.35	8.35	8.35
Kh = Inhibitor Constant (Lb-°F/lbmol)	2,335	2,335	2,335	2,335	2,335
MW = Methanol Mol. Wgt. (Lb/mol)	32.042	32.042	32.042	32.042	32.042
= Req'd. Weight % Inhibitor In Liquid Water Phase	10.28	10.28	10.28	10.28	10.28
Wv = MeOH Lost To Vapor Phase (Lb/Day)	310	310	310	310	310
Wl = MeOH In Water Liquid Phase (Lb/Day)	2,527	2,527	2,527	2,527	2,527
Wc = MeOH In Condensate Liquid Phase (Lb/Day)	6,406	3,154	2,092	1,670	1,111

Output Values:

Th = Gas Hydrate Temperature (°F)		68.3	68.3	68.3	68.3	68.3
Wm = Req'd. Methanol Mass Flow Rate	(Lb/Day)	9,243	5,991	4,929	4,507	3,948
	(Lb/Hr)	385	250	205	188	164
	(Lb/MMscf)	462	300	246	225	197
Qm = Req'd. Methanol Liquid Volume Flow Rate	(Gal/Day)	1,401	908	747	683	598
	(BBL/Day)	33.3	21.6	17.8	16.3	14.2
	(Gal/Min)	0.97	0.63	0.52	0.47	0.42
	(Gal/MMscf)	70.07	45.39	37.35	34.15	29.91
	(BBL/MMscf)	1.67	1.08	0.89	0.81	0.71

FIGURE 1-25 Spreadsheet illustrating the sensitivity of the solubility of methanol.

Note the sensitivity of the total methanol requirements as the solubility of methanol is varied from 0.5 wt.% to 3.0 wt.% in the following spreadsheet for our example (Figure 1.25). (Current research reports and laboratory analysis suggest the solubility is actually closer to 0.5 wt.%.)

EXERCISES

1. Calculate the water content of the following gas stream at 2000 psia and 100°F.

Component	Mol %
N_2	8.5
H_2S	5.4
CO_2	0.5
C_1	77.6
C_2	5.8
C_3	1.9
iC_4	0.1
nC_4	0.1
iC_5	0.1
	100.0

Use the following:

(a) McKetta-Wehe graph

(b) Weighed average method (Equation 1-2)

(c) SRK "quick look" method

2. A gas stream, saturated with water, leaves the wellhead at 122°F and 2900 psia. Some distance downstream the gas enters a separator at 1015 psia and 50°F. How much liquid water should be drained from the separator, if any?

3. Given a gas stream at 1000 psia, MW = 20.37 and the following composition:

Component	Mol %
N_2	10.1
C_1	77.1
C_2	6.1
C_3	3.5
iC_4	0.7
nC_4	1.1
C_{5+}	0.8
	100.0

Determine the hydrate formation temperature using:

(a) Vapor-solid equilibrium constant

(b) Pressure-temperature correlation curves

4. Determine the temperature drop across a choke with a flowing tubing pressure of 5000 psia and a downstream back pressure of 1000 psia. The well stream produces 60 bbl/MMSCFD of liquid hydrocarbon.

5. 9.5 MMSCFD of natural gas (S = 0.65) having a hydrate
    formation temperature of 70°F cools to 40°F in a burned
    pipeline. Assume pipeline pressure is 900 psia. How
    much methanol must be added in bbl/day, if the gas
    enters the line saturated at 90°F and is free of liquid
    water.

Part 2
Dehydration Considerations

Contents

OVERVIEW

If hydrate prevention methods are unsuitable and hydrates are liable to form, some water must be removed from the gas stream.

Dehydration is the process of removing water from the stream.

Water removal from gas can be accomplished by several processes, the two most common methods are:

Adsorption

Absorption

One less common method of dehydration will also be described, and that is:

Nonregenerable dehydrator (calcium chloride brine unit)

DOI: 10.1016/B978-1-85617-980-5.00002-1

ADSORPTION

Process Overview

Adsorption is a physical phenomenon that occurs when molecules of a gas are brought into contact with a solid surface and some of them condense on the surface.

Dehydration of a gas (or a liquid hydrocarbon) with a dry desiccant is an adsorption process in which water molecules are preferentially held by the desiccant and removed from the gas stream.

Adsorption involves a form of adhesion between the surface of the solid desiccant and the water vapor in the gas.

Water forms a thin film that is held to the desiccant surface by forces of attraction, not by chemical reaction.

Desiccant is a solid, granulated dehydrating medium with a large effective surface area (large number of small pores) per unit weight.

Typical desiccants might have as much as 4 million square feet of surface area per pound.

Commonly used desiccants include:

> Alumina
>
> Silica gel
>
> Molecular sieves (mol sieves)

Many grades and qualities of each of these substances are commercially available.

Figure 2-1 is an enlargement of a molecular sieve particle.

Principles of Adsorption

The achievement of equilibrium on a small surface displays the following pattern:

> Some passing molecules will condense on the surface (physical as opposed to chemical absorption).
>
> After some finite time the molecule may acquire sufficient energy to leave and be replaced by another.
>
> After sufficient time, a state of equilibrium will be reached wherein the number of molecules leaving the surface will equal the number arriving.

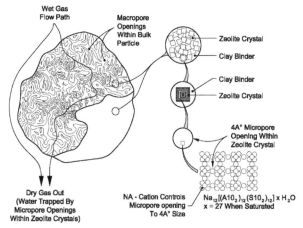

FIGURE 2-1 Enlargement of a molecular sieve particle.

The number of molecules on the surface is a function of:

The nature of the adsorbent

The nature of the molecule being adsorbed (the adsorbate)

The temperature of the system and concentration of the adsorbate over the adsorbent surface

Process Reversal

The adsorption process may be reversed in the same manner that absorption processes are.

Adsorption is encouraged by low temperatures and high pressures.

Desorption (its reversal) is encouraged by high temperatures and low pressures.

Mass Transfer Zone (MTZ)

At the inlet of the bed and for a certain distance into it, the adsorbent is saturated to essentially equilibrium value with the adsorbable component in the fluid, such as water in natural gas.

At the outlet of the bed, the adsorbent is unsaturated and the water content of the gas is in equilibrium with the unsaturated activated adsorbent.

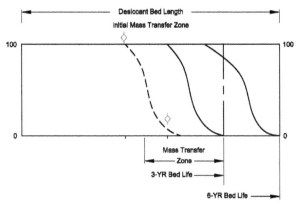

FIGURE 2-2 Schematic view of Mass Transfer Zone (MTZ).

The MTZ is defined as the zone between these two zones, where the concentration of the water in the natural gas is falling (Figure 2-2).

MTZ lengths can be obtained experimentally for various materials and systems and used in graphical correlations for design purposes.

MTZ is a function of the following factors:

Adsorbent

Adsorbent particle size

Fluid velocity

Fluid properties

Adsorbate concentration in the entering fluid

Adsorbate concentration in the adsorbent if it is not fully reactivated

Temperature

Pressure

Past history of the system

Principles of Operation

Introduction

The adsorption process is a batch process, with multiple desiccant beds used in cyclic operation to dry the gas on a continuous basis.

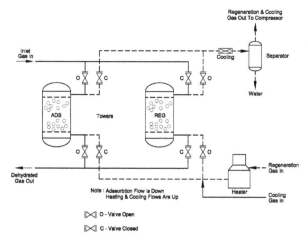

FIGURE 2-3 Simplified flow diagram of a two-tower solid desiccant dehydration system.

The number and arrangement of the desiccant beds may vary from two towers, adsorbing alternatively (Figure 2-3), to many towers.

Three separate functions or cycles must alternatively be performed in each dehydrator tower:

Adsorbing or gas-drying cycle

Heating or regeneration cycle

Cooling cycle (prepares the regenerated bed for another adsorbing or gas-drying cycle)

Figure 2-4 is a flow diagram of a typical two-tower dehydration unit.

System Components

Essential components of a solid desiccant dehydration system are:

Inlet gas stream microfiber filter separator

Two or more adsorption towers (contactors) filled with a solid desiccant

High-temperature heater to provide hot regeneration gas to reactivate the desiccant in the towers

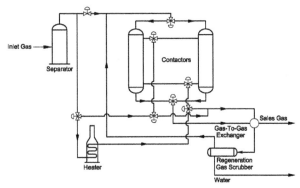

FIGURE 2-4 Flow diagram of a solid desiccant dehydration unit.

Regeneration gas cooler to condense water from the hot regeneration gas

Regeneration gas separator to remove the condensed water from the regeneration gas

Piping manifolds, switching valves, and controls to direct and control the flow of gases according to the process requirements

Drying/Reactivation Cycle

Figure 2-5 shows the flow of a typical two-tower unit with drying taking place in the first tower.

Wet inlet gas first passes through an efficient microfiber inlet filter separator where free liquids, entrained mist, and solid particles are removed.

Free liquids may damage or destroy the desiccant bed.

Solids may plug the bed.

If the dehydration unit is downstream of an amine unit, glycol unit, or compressors, a microfiber filter inlet separator is highly recommended upstream of the adsorber towers.

At any given time, one of the towers will be on stream in the adsorbing or drying cycle while the other is in the process of being heated or cooled.

Several automatically operated switching valves and a controller route the inlet gas and regeneration gas to the right tower at the proper time.

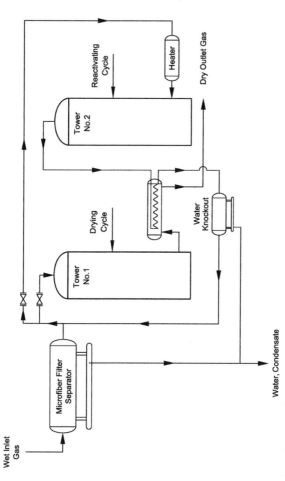

FIGURE 2-5 Flow diagram of a typical two-tower natural gas dehydrator.

The tower being regenerated is:

> Heated for 5 to 6 hours

> Cooled for the remaining 2 to 3 hours

As the wet gas flows downward through the tower on the adsorption cycle, each of the adsorbable components is adsorbed at a different rate.

The water vapor is immediately adsorbed in the top layers of the desiccant bed.

Some of the light hydrocarbon gases and heavier hydrocarbons moving down through the bed are also adsorbed.

Heavier hydrocarbons will displace the lighter ones in the desiccant bed as the adsorbing cycle proceeds.

As the upper layers of desiccant become saturated with water, water in the wet gas stream begins displacing the previously adsorbed hydrocarbons in the lower layers.

For each component in the inlet gas stream, there will be a section of bed depth, from top to bottom, where the desiccant is saturated with that component and where the desiccant below is just starting to adsorb it.

The depth of bed from saturation to initial adsorption is the mass transfer zone (MTZ).

MTZ is simply a zone or section of the bed where a component is transferring its mass from the gas stream to the surface of the desiccant.

As the flow of gas continues:

> The MTZs move downward through the bed and water displaces all of the previously adsorbed gas until, finally, the entire bed is saturated with water vapor.

When the bed is completely saturated with water vapor, the outlet gas is just as wet as the inlet gas.

Towers must be switched from the adsorption cycle to the regeneration cycle (heating and cooling) before the desiccant bed has become completely saturated.

One regeneration-gas supply scheme consists of taking a portion (5 to 15%) of the entering wet gas stream across a pressure-reducing valve that forces a portion of the upstream gas through the regeneration system.

In most plants, a flow controller regulates the volume of regeneration gas used.

Regeneration gas is sent to a heater where it is heated to between 400°F and 600°F and then piped to the tower being regenerated.

Initially, the hot regeneration gas must heat up the tower and the desiccant.

The water begins vaporizing when the effluent hot gas temperature reaches between 240°F and 250°F.

The bed continues to heat up slowly as the water is being desorbed or driven out of the desiccant.

After all the water has been removed, heating is maintained to drive off any heavier hydrocarbons and contaminants that would not vaporize at lower temperatures.

The desiccant bed will be properly regenerated when the outlet gas (peak-out) temperature has reached between 350°F and 550°F.

After the heating cycle, the desiccant bed is cooled by flowing unheated regeneration gas until the desiccant is sufficiently cooled.

All of the regeneration gas used in the heating and cooling cycles is passed through a heat exchanger (normally an aerial cooler) where it is cooled to condense the water removed from the regenerated desiccant bed.

This water is separated in the regeneration gas separator, and the gas is mixed with the incoming wet gas stream.

This entire procedure is continuous and automatic.

Performance

Advantages

Can achieve very low dew points (less than 1 ppm)

High contact temperatures are possible

Adaptable to large rate and load changes

Disadvantages

High initial cost

Batch process

Experiences high-pressure drop through the bed

Desiccant is sensitive to poisoning with liquids or other impurities in the gas

Effect of Process Variables

Several process variables can have a major effect on dry bed dehydration sizing and operating efficiency:

Quality of inlet gas

Temperature

Pressure

Cycle time

Gas velocities

Sources of regeneration gas

Desiccant selection

Effect of regeneration gas on outlet gas quality

Pressure drop considerations

Quality of Inlet Gas

Performance of dry bed dehydrator is affected by:

Moisture content of inlet gas

Components in the produced natural gas stream

The relative saturation of the inlet gas:

Determines the size of a given desiccant bed

Affects the transfer of water to the adsorbent

Higher capacities can be expected when drying saturated gas (100% relative humidity) for most desiccants (except molecular sieve) then when drying partially saturated gases.

In most gas field applications, the inlet gas is saturated with water vapor and thus this variable need not be considered.

Compounds in produced natural gas adversely affect performance of the dry bed dehydrator.

Components of concern are:

Carbon dioxide

Heavy hydrocarbons

Sulfur-bearing compounds

The greater the molecular weight of a compound, the greater its adsorption potential.

Temperature
General Considerations

Operation is very sensitive to the temperature of the incoming gas.

Efficiency decreases as the temperature increases.

Molecular sieves and most other adsorbents have significantly higher adsorptive capacity at low temperatures.

Figure 2-6 indicates this characteristic for both silica gel and a type 5A molecular sieve.

Water capacity of silica gel at 80°F increases to over twice that for the molecular sieve at higher water partial pressures.

Temperature of the regeneration gas that commingles with the incoming wet gas ahead of the dehydrators is important.

The temperature must remain within 10°F to 15°F, otherwise liquid water and hydrocarbons will condense as the hotter gas stream cools.

Condensed liquids that strike the bed can shorten the solid desiccant's life.

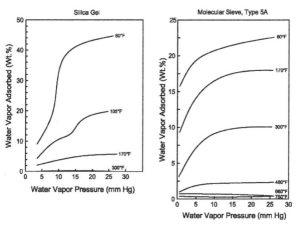

FIGURE 2-6 Effect of temperature on adsorption capacity.

Temperature of the hot gas entering and leaving a desiccant tower during the heating cycle affects plant efficiency and the desiccant life.

High regeneration gas temperature assures good removal (desorption) of water and contaminants from the bed.

Maximum Temperature

The maximum temperature depends upon:

The type of contaminant

"Holding power" or affinity of the desiccant for the contaminants

Typically 450°F to 600°F is used.

Desiccant bed temperature reached during the cooling cycle is important.

If wet gas is used to cool the desiccant:

Terminate the cooling cycle when the bed reaches 125°F.

Additional cooling may cause water to be adsorbed from the wet gas stream and preload (presaturate) the bed before the next adsorption cycle begins.

If dry gas is used to cool the desiccant:

Terminate the cooling cycle within 10°F to 20°F of the incoming gas temperature.

It maximizes adsorption capacity of the bed.

The temperature of the regeneration gas going through the regeneration gas scrubber should be held low enough to condense and remove the water and hydrocarbons without causing hydrate problems.

Pressure

The adsorption capacity of a dry bed unit decreases as pressure is lowered and with usage.

Operating dry bed dehydrators well below the design pressure requires the desiccant to work harder to:

Remove the additional water

Maintain the desired effluent dew point

With the same volume of incoming gas, the increased gas velocity occurring at the lower pressure could:

> Affect the effluent moisture content

> Damage the desiccant

At pressure above 1300 to 1400 psia, the co-adsorption effects of hydrocarbons are very significant.

Cycle Time

Most adsorbers operate on a fixed drying cycle time which is frequently set for the worst conditions.

Adsorbent capacity is not a fixed value and declines with usage.

For the first few months of operation, a new desiccant normally has a high capacity for water removal.

If a moisture analyzer is used on the effluent gas, a much longer drying cycle can be achieved.

As the desiccant ages, the cycle time can be shortened to save regeneration fuel costs and improve the desiccant life.

Common cycle times

> 8 hours on stream

> 5 to 6 hours heating

> 2 to 3 hours cooling

Gas Velocities

As the gas velocity during the drying cycle decreases, the ability of the desiccant to dehydrate the gas increases by:

> Lowering effluent moisture contents

> Longer drying cycle times

Figure 2-7 shows the general effect of gas rate on the extent of dehydration.

On the surface, it would seem desirable to operate at minimum flow rates to utilize the desiccant fully. However, low linear velocities:

> Require towers with large cross-sectional areas to handle a given gas flow

> Allow wet gas to channel through the desiccant bed and thus not be properly dehydrated

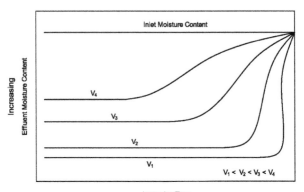

FIGURE 2-7 Series of adsorption curves that show effects of flow rate on the adsorption capacity of a single desiccant.

Compromise must be made between the tower diameter and the maximum utilization of the desiccant, as illustrated in Figure 2-8. Maximum superficial velocities are shown in Table 2-1.

High linear velocities:

> Lower adsorption efficiency

> May cause desiccant damage

Minimum tower diameter can be determined from the following:

$$d^2 = 3600 \left(\frac{Q_g T Z}{V P} \right) \qquad (2\text{-}1)$$

Where:

> d = Tower internal diameter, in.
>
> Q_g = Gas flow rate, MMSCFD
>
> T = Gas temperature, °R
>
> Z = Compressibility factor
>
> V = Gas superficial velocity, ft./min (Table 2-1)
>
> P = Tower operating pressure, psia

The regeneration gas velocity is important, especially when effluent moisture contents below 1 ppm are needed.

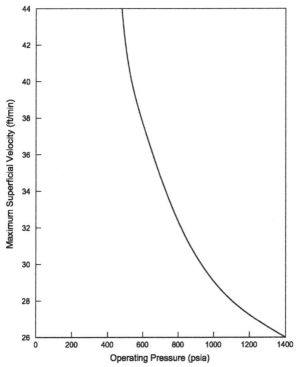

FIGURE 2-8 Maximum downflow gas velocity as a function of operating pressure.

Table 2-1 Maximum Superficial Velocities

Tower Operating Pressure (psig)	Max. Superficial Velocity (ft./min)
14.7	110
400	60
600	55
1000	40

At velocities less than 10 ft./sec., hot gas will channel through the bed, leaving excess water in the bed after regeneration which results in poor dehydration.

Source of Regeneration Gas

Source of regeneration gas depends on:

Plant requirements

Availability of a suitable gas stream

Regeneration gas should be dry when low effluent moisture contents (in the range of 0.1 ppm) are required.

Plant tail gate gas can normally be used.

If only moderate drying is required, a portion of the wet feed gas can be used.

Figure 2-9 is an equilibrium diagram showing lines of constant water loading. For example:

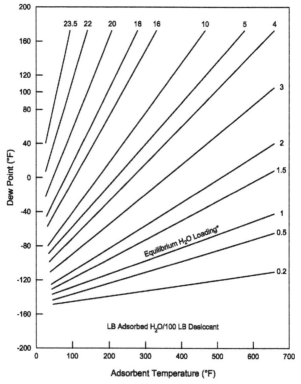

FIGURE 2-9 Equilibrium diagram showing lines of constant water loading for a type 4a molecular sieve.

A molecular sieve bed at 100°F in equilibrium with a gas having a −80°F water dew point will contain about 4 wt.% water.

Equilibrium curves for a given adsorbate-adsorbent can be used to estimate the regeneration conditions necessary to provide the required outlet conditions. For example,

If the regeneration gas is taken from inlet gas with a dew point of 40°F and is heated to 450°F, the mol sieve will contain 3 wt.% water after regeneration.

If the gas to be treated is at 100°F, the intersection of the 3 wt.% line with an adsorbent temperature of 100°F gives the minimum attainable dew point at −95°F.

If this dew point is not satisfactory, either the regeneration gas must be heated to above 450°F or a gas of a higher dew point (e.g., residue gas) must be used for regeneration gas.

Direction of Gas Flow

Flow direction influences:

Effluent purity

Regeneration gas requirements

Desiccant life

Direction of flow during the drying cycle is downward, which:

Permits higher velocities without lifting or fluidizing the desiccant bed

Means fluidization can severely damage the desiccant

Direction of flow during the heating cycle is counter-current to the direction of the adsorption flow.

It permits better reactivation of the lower portion of the desiccant bed, which must perform the super-dehydration during the drying cycle, especially in cryogenic plants.

If flow is cocurrent, all water and/or other contaminants must move through the entire bed, thus causing additional desiccant contamination and requiring longer regeneration times.

Direction of flow during the cooling cycle:

> When dry gas is used, the flow direction is counter current to the adsorption flow, thus simplifying piping and valve configuration.

> When wet gas is used, the flow direction is in the same direction as the adsorption flow so that the water adsorbed during the cooling cycle as the desiccant cools will preload on the inlet end of the bed.

If counter current flow is used, water is deposited on the exit end of the bed:

> When the next adsorption cycle begins, the wet gas is immediately dried.

> As the dry gas continues to move down through the bed, it picks up some of the water deposited during the cooling cycle and sometimes puts too much moisture in the effluent stream.

If wet gas is used, the additional water load, deposited during the cooling cycle, should be included when the amount of desiccant needed for dehydration is calculated.

Desiccant Selection

No desiccant is best for all applications.

Desiccant selection is based upon:

> Economics

> Process conditions

Desiccants are usually interchangeable.

Equipment designed for one desiccant can often operate effectively with another.

No desiccant product will remain effective with massive liquid carryovers.

All desiccants are damaged by heavy impurities carried into the bed with gases. These include:

> Crude oil and condensate

> Glycols and amines

> Most corrosion inhibitors

> Well treating fluids

All desiccants exhibit a decrease in capacity (design loading) with an increase in temperature

Molecular sieves are less affected.

Aluminas are most affected.

Aluminas and molecular sieves act as a catalyst with H_2S to form COS, which deposits sulfur on the desiccant bed during regeneration.

Alumina gels, activated aluminas, and molecular sieves are all chemically attacked by strong mineral acids and thus decrease their adsorptive capacity.

Special acid resistant molecular sieves desiccants are available.

Table 2-2 provides certain physical characteristics of the more common solid desiccants.

Molecular Sieves

Offer the highest adsorptive capacity of all desiccants when the feed gas is at very high temperatures or at low relative saturation.

Only desiccants capable of dehydrating gas to less than 1 ppm of water content are required for cryogenic temperatures (dew points down to $-150°F$).

Silica Gel and Alumina

Water saturated gases entering the dehydrator can adsorb twice as much water as molecular sieves and offer a lower first cost.

Table 2-2 Properties of Solid Desiccants

Desiccant	Bulk Density (lb/ft.^{3})	Specific Heat (Btu/lb/°F)	Normal Sizes Used	Design Adsorptive Capacity (wt.%)
Activated Alumina	51	0.24	¼ in.–8 mesh	7
Mobil SOR Beads	49	0.25	4–8 mesh	6
Fluorite	50	0.24	4–8 mesh	4–5
Alumina Gel (H-151)	52	0.24	⅛–¼ inch	7
Silica Gel	45	0.22	4–8 mesh	7
Molecular Sieves (4A)	45	0.25	⅛ inch	14

Silica Gel

Silica gel can be regenerated to a lower water content than molecular sieves and at much lower temperatures (400°F for gels versus 500° to 600°F for sieves).

It shatters in the presence of free water or light hydrocarbon liquids.

The problem is minimized by using a 4 to 6 inch buffer bed of mullite ball (or equivalent) to protect the silica gel from direct contact.

Desirable Characteristics of Solid Desiccants

High adsorptive capacity (lb/lb), which reduces contactor size

Easy regeneration, for simplicity and economics of operation

High rate of adsorption, which allows higher gas velocities and thereby reduces contactor size

Low resistance to gas flow, to minimize gas pressure drop through the unit

High adsorptive capacity retained after repeated regeneration, allowing smaller initial charge and longer service before replacement

High mechanical strength, to resist crushing and dust formation

Inert chemicals, to prevent chemical reactions during adsorption and regeneration

Volume unchanged when product is wet, which would otherwise necessitate costly allowance for expansion

Noncorrosive and nontoxic properties, eliminating the necessity for special alloys and costly measures to protect the operator's safety

Low cost, to reduce initial and replacement costs

Effect of Regeneration Gas on Outlet Gas Quality

Regeneration gas desorbs molecular sieve beds chromatographically in the reserve order of the adsorption bead. For example:

Adsorbed methane and ethane would be desorbed first, then propanes and heavier hydrocarbons, then carbon dioxide, followed by any hydrogen sulfide that might have been in the inlet gas, and last of all, the water.

The effect of the concentration of these impurities in the regeneration gas stream may be significant when regeneration gas is 10 to 15% of the net inlet gas.

In the regeneration circuit, the bulk of the water and some heavy hydrocarbons are condensed and removed from the system:

They may render the sales gas off-specification for a short period.

The peak of ethane or CO_2 could cause the sales gas to exceed its heating value.

Concentrations of 3 to 4 ppm of H_2S can be concentrated up to 20 times that amount, and thus render the composite stream far off-spec.

Figure 2-3 shows the cooled regeneration gas stream is recombined with the main gas inlet to be processed.

This recycle essentially eliminates the problem of making the sales gas off-specification.

But it adds cost to the extent that the main gas processing capacity must be increased appropriately.

If the sales gas limits are no problem, or if there is other downstream processing, the cooled, scrubbed regeneration gas may be admitted directly to the dried outlet gas without this recycle.

Pressure Drop Considerations

To achieve acceptable dehydration and extend the life of the desiccant, the pressure drop through the dehydration tower should not exceed 8 psi.

Pressure drop through the tower can be estimated from either:

Desiccant pressure drop curves furnished by the manufacturer (Figure 2-10), or

Pressure drop equation

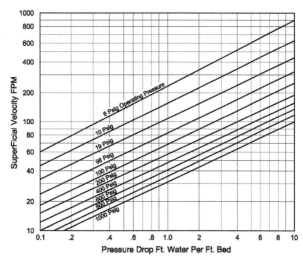

FIGURE 2-10 Typical pressure-drop curve for silica gel type desiccants, 0.15-inch diameter beads.

Pressure drop through a desiccant tower can be estimated from the following equation:

$$\frac{\Delta P}{L} = B\mu V + C\rho V^2 \qquad (2\text{-}2)$$

Where:

ΔP = Pressure drop through the tower, psi (typically sized for 5 psi)

μ = Gas viscosity, cp

ρ = Gas density, lb/ft.^{3}

V = Gas superficial velocity, ft./min

B, C = Constants provided in Table 2-3

Table 2-3 Constants Used in Pressure Drop Equations

Particle Type	B	C
⅛-inch bead	0.0560	0.0000889
⅛-inch extruded	0.0722	0.0001240
¹⁄₁₆-inch bead	0.1520	0.0001360
¹⁄₁₆-inch extruded	0.2380	0.0002100

Example 2-1: Determination of Pressure Drop through a Dry Bed Desiccant Dehydration Tower

Given:

Superficial gas velocity = 40 ft./min

Tower operating pressure = 1000 psig

Gas molecular weight = 18

Bed height (L) = 30 ft.

Desiccant type = Silica gel

Desiccant diameter = 0.15 inch

Desiccant pressure drop curve (Figure 2-10)

Notes:

1. Curves are based on air flow. For other gases, multiply pressure drop by

$$\left(\frac{MW_{Gas}}{MW_{Air}}\right)^{0.9} \qquad (2\text{-}3)$$

2. Pressure drop curves are based on clean beds. After about 2 years, the beds will foul somewhat and the pressure drop will be about 1.6 times the value read from the curves.

Solution:

1. Enter Figure 2-10, extend horizontal line from superficial velocity of 40 ft./min, and intersect with the operating pressure of 1000 psig.

2. Draw a vertical straight line down from the intersection and read a pressure drop of 1.9 feet of water per foot of bed.

3. Calculate the total pressure drop across the bed after two years of service,

$$\text{Total } \Delta P = \left(1.9 \frac{\text{Ft. of H}_2\text{O}}{\text{Ft. of Bed}}\right)\left(0.433 \frac{\text{psi}}{\text{Ft. of H}_2\text{O}}\right) x$$

$$\left(\frac{18}{29}\right)^{0.9}(1.6)(30\,\text{ft.})$$

$$= 25\,\text{psi}$$

Equipment

The proper selection of equipment is essential to good operations.

Inlet Gas Cleaning Equipment

All hydrocarbon liquids, free water, glycol, amine, or lube oil carry over must be cleaned from the inlet gas to ensure the best dry desiccant dehydrator operation.

In all cases, the dry bed unit should have a scrubber (or a filter separator) between it and a primary well fluid separator.

A microfiber filter separator (or its equivalent) should always be installed upstream of the inlet scrubber if a carryover of glycols, amines, or compressor lube oils is possible.

Liquid level controls need to be checked frequently as well as the liquid dump line to ensure their operability.

Adsorber Tower
General Considerations

An adsorber is a cylindrical tower filled with a solid desiccant.

Depth of desiccant will vary from a few feet to 30 feet or more.

Vessel diameter may be as much as 10 to 15 feet or more.

Bed height to diameter (L/D) ratio of 2.5–4.0 to 1 is desirable.

Lower ratios (1:1) are sometimes used, which could result in poor gas dehydration caused by:

Non-uniform flow

Channeling

Inadequate contact time between the wet gas and the desiccant

Three problems that frequently cause poor operation are:

Insufficient gas distribution

Inadequate insulation

Improper bed supports

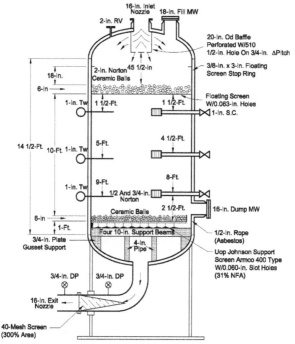

FIGURE 2-11 Molecular sieve gas dehydration tower.

Figure 2-11 illustrates many of the desired features recommended in a dry bed desiccant tower.

Insufficient Gas Distribution

Poor gas distribution at the inlet and outlet of the desiccant beds has caused many costly problems, resulting in:

Channeling

Desiccant damage

The inlet gas distributor should be provided with adequate baffling before the gas enters the desiccant bed.

Void space of 18 to 24 inches is recommended.

Neither gas to be dehydrated nor the regeneration gas should impinge directly on the bed.

Channeling, high localized velocities and swirling can cause:

Desiccant attrition

High-pressure drop through the desiccant bed as attrition fines lodge between the regular particles

Screen-wrapped slotted pipe, with gas at low velocities exiting radially into the vessel is recommended.

A 4- to 6-inch layer of large diameter (2 inch) support balls can be placed on top of the desiccant bed. This:

Improves gas distribution

Prevents desiccant damage from swirling

Swirling can destroy several feet of castable refractory lining by turning the powdered desiccant into a sandblasting agent which results in:

High heat losses

Poor desiccant regeneration

Inadequate Insulation

Internal or external insulation can be used.

Internal insulation:

Reduces the total regeneration gas requirements and costs

Eliminates the need to heat and cool adsorber vessels

Provision must be made for expansion and contraction so that there will be no cracking or weld failures:

It is normally made from a castable refractory lining.

Liner cracks permit some of the wet gas to bypass the desiccant bed.

Only a small amount of wet, bypass gas can cause freeze up in cryogenic plants.

Ledges installed every few feet along the vessel wall can help eliminate liner cracks.

Improper Bed Supports

Two common bed supports include:

> Horizontal screen supported by I-beams and a welding ring

> Vessel whose bottom head is filled with graduated support balls

Screens are usually made of stainless steel or monel that have openings at least 10 meshes smaller than the smallest desiccant particle:

> 0.033-inch slot opening will retain standard desiccant particles.

> Screen traps crush desiccant particles, which prevents the malfunction of downstream equipment caused by invasion of these particles as the desiccant deteriorates.

Screens should be securely fastened in the vessel.

Provisions should be made for expansion and contraction as the adsorbers heat and cool.

Annular space between the vessel wall and the edge of the bed support screen must be sealed to prevent the loss of desiccant:

> Asbestos rope packing, forced in this space, is used.

> A support ring around the edges of the screen is beneficial.

If the screens are installed in sections, they should be fastened securely with stainless steel wire.

Support balls on the screens are helpful.

2 to 3 inches of ½-inch balls are gently placed on the screen and a 2 or 3 inch smooth layer of ¼-inch balls is gently placed on top of the ½-inch balls.

> These layers prevent desiccant dust or whole particles from plugging the screen openings and forcing a high-pressure drop across the desiccant beds.

When calculating the regeneration needs of the system, it is important to include the heat requirements for the support balls.

If the bottom head of the vessel is filled with graduated support balls, a gas distributor may be required between the balls and the lower portion of the desiccant bed when upflow heating or cooling is used.

This is important on large-diameter vessels to prevent channeling and poor reactivation of the desiccant.

Many adsorbers have a void area in the bottom, below the bed supports, to collect contaminants, dust, and fines.

A blowdown nozzle can be provided to discharge these materials.

A moisture sample probe should be located in the adsorbers in cryogenic plants several feet from the outlet end of the bed and extending to the center.

This probe, used in conjunction with the outlet gas moisture probe, offers valuable flexibility in studying and solving dehydrator problems, particularly for determining if gas is being channeled down the walls of the vessel.

It permits capacity tests for optimizing drying cycle times.

Tests can be conducted with reasonable safety because movement of the waterfront can be detected prior to breakthrough.

Probe can be a long thermowell drilled with $\frac{1}{32}$-inch holes on the sides near the end of the probe.

Pressurization

For best performance and maintenance of desiccant quality, adsorbers should:

Never be pressurized faster than 50 psi/min

Never be depressurized faster than 10 psi/min

Downflow pressure drop should not exceed 1 psi/ft.

Upflow pressure drop should not be less than $\frac{1}{4}$ psi/ft. to prevent fluff fluidization

Even with the best designs, some desiccant dust is swept out of the beds at design gas-flow rates.

Certain amounts can be tolerated in many field dehydration systems.

It is not acceptable in turbo expander plant designs that involve extensive downstream heat exchange and processing.

The problem is particularly significant where plate-fin or core-type heat exchangers are used.

In many instances, this problem can be solved with microfiber filters (cleaning to 1 micron) with a differential pressure across them of 15 psi.

Regeneration Gas Exchangers, Heaters, and Coolers

A gas or gas exchanger is usually designed with the following assumptions:

All of the water will be liberated from the bed in 1 hour at 250°F.

Regeneration gas can be cooled to within 10°F of the sales gas temperature.

A regenerative gas heater is sized to provide:

Heat to desorb the water

Heat for the desiccant of between 500° and 550°F

Heat the contactor shell

The heat of desorption:

For silica gel is 1100 Btu/lb of water

For a molecular sieve is approximately 50% higher

Heat required to heat the desiccant can be calculated by using the following equation

$$Q = Wc_p\Delta t \tag{2-4}$$

Where:

Q = Heat required, Btu

W = Weight of desiccant, lb

C_p = Specific heat of desiccant, Btu/lb/°F

Δt = Difference in desired bed temperature and normal bed operating temperature, °F

Sensible heat for the contactor shell can be calculated using Equation 2-4 and by:

Estimating the weight of the steel

Using 0.12 Btu/lb/°F for c_p

On units that have internal insulation, the heat transferred to the shell is considered negligible.

Normal practice is to add 10 to 20% to the sum of the heat required in order to account for heat losses and add some margin of safety.

Regeneration Gas Separator

Most desiccants also have an affinity for hydrocarbons, thus a skimmer is used to separate the valuable hydrocarbons from the water to be discarded.

Frequent pH tests on the discarded water helps pinpoint corrosion problems in the adsorption system.

A common problem encountered in regeneration gas separators is the fouling of the liquid dump line by desiccant dust and heavy oils.

It allows liquids to be carried back where they can damage the desiccant in a recycle system or contaminate sales and downstream facilities in a once-through system.

Regular inspection and cleaning are required to prevent such damage.

Control Valves

Quality valves should be used to prevent costly operating problems.

Generally, two-way valves have fewer problems than three-way valves.

Most difficult service is encountered where the valves have hot regeneration gas (600°F) on one side and ambient (100°F) gas on the other.

Careful piping design can reduce this large gradient.

Valve sequencing is important to prevent a sudden upflow caused by a pressure difference.

The problem could fluidize the bed and damage the desiccant.

Dry bed dehydrators equipped with motor valves for switching operations require frequent servicing to eliminate leakage.

Expander Plant Molecular Sieve Applications

Turbo expander plants commonly operate down to temperatures of −150°F.

Operating points:

> Much below the equilibrium water content data illustrated in McKetta-Wehe chart

Include designs to water contents as low as 1 ppm.

As shown in Table 2-2, only molecular sieves and activated alumina are capable of such performance.

Molecular sieves are used in approximately 95% of the dehydration equipment for this type of plant (a 4A molecular sieve has twice the adsorptive capacity of activated alumina).

Figure 2-12 compares the adsorptive capacity of several desiccants at lower relative humidities of the gas.

At 30% relative humidity, molecular sieves would adsorb 21.5 lb water per 100 lb of desiccant, whereas silica gel would absorb 15 lb of water/ 100 lb of desiccant.

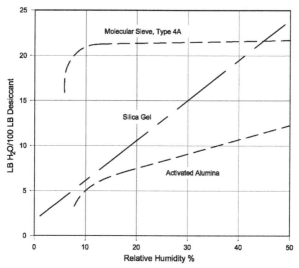

FIGURE 2-12 Water vapor adsorption at 60°F.

Desiccant Performance

General Conditions

Desiccants decline in adsorptive capacity at different rates under varying operating conditions.

Desiccant aging is a function of many factors, including

Number of cycles experienced

Exposure to any harmful contaminants present in the inlet stream that are not completely removed during normal reactivation.

The single most important variable affecting the decline rate of desiccant capacity is the chemical composition of the gas or liquid to be dried.

Feed stream composition should always include the contaminants.

Capacity of a new desiccant will decline slowly during the first few months in service because of cyclic heating, cooling, and netting.

Desiccant capacity usually stabilizes at about 55 to 70% of the initial capacity.

Moisture Analyzer

Used to optimize the drying cycle time.

Allows drying time to be shortened as the desiccant ages.

Both inlet and outlet moisture analyzer probes should be used.

A probe extending approximately 2 feet upward into the bed from the outlet end is recommended because it allows a dehydration capacity test to be run without the risk of a water breakthrough.

Effect of Contaminants in Inlet Feed Stream

Compressor oils, corrosion inhibitors, glycols, amines, and other high-boiling contaminants cause a decline in desiccant capacity, because normal reactivation temperatures will not vaporize the heavy materials.

Residual contaminants slowly build up on the desiccant's surface reducing the area available for adsorption.

Many corrosion inhibitors chemically attack certain desiccants, permanently destroying their usefulness.

Effect of Regeneration Gases Rich in Heavy Hydrocarbons

Use of this rich gas in a 550° to 600°F regeneration service aggravates coking problems.

Rich gases may be dried satisfactorily with molecular sieves.

Lean dry gas is always preferable for regeneration.

Effect of Methanol in the Inlet Gas Stream

Methanol in the inlet gas is a major contributor to the coking of molecular sieves where regeneration is carried out at temperatures above 550°F.

Polymerization of methanol during regeneration produces dimethyl ether and other intermediates that will cause coking of the beds.

Conversion to ethylene glycol injection, instead of methanol for hydrate control, will increase sieve life and add at least 10% to sieve capacity by removing the vapor phase methanol from the system.

Useful Life

Ranges from one to four years in normal service.

Longer life is possible if feed gas is kept clean.

Effectiveness of regeneration plays a major role in retarding the decline of a desiccant's adsorptive capacity and prolonging its useful life.

If all the water is not removed from the desiccant during each regeneration, its usefulness will sharply decrease.

Effect of Insufficient Reactivation

Insufficient reactivation can occur if the regeneration gas temperature or velocity is too low.

A desiccant manufacturer will generally recommend the optimum regeneration temperature and velocity for the product.

Velocity should be high enough to remove the water and other contaminants quickly, thus minimizing the amount or residual water and protect the desiccant.

Effect of High Reactivation Temperature

Higher reactivation temperatures remove volatile contaminants before they form coke on the desiccant.

Maximizes desiccant capacity and ensures minimum effluent moisture content.

Final effluent hot gas temperature should be held one or two hours to achieve effective desiccant reactivation.

Areas Requiring Engineering Attention

The design of adsorption type dehydration equipment can be improved by considering the following:

The major process variables that effect bed loading.

Proper conditioning of inlet gas and proper design of regeneration gas systems.

An accurate estimation of bed sizes in order to realistically evaluate competitive offering of desiccant vendors.

Improvement in the design of adsorber internals including internal insulation, improved switching valves, and control systems.

Example 2-2: Preliminary Solid Bed Desiccant Design

Note: Detailed design of dry bed dehydrators should be left to experts. The general "rule of thumb" presented herein can be used for preliminary design.

Given:

Feed rate	= 50 MMSCFD
Molecular weight of gas	= 17.4
Operating temperature	= 110°F
Operating pressure	= 600 psia
Inlet dew point	= 100°F (equivalent to 90 lb H_2O/MMSCF)
Desired dew point	= 1 ppm H_2O
Gas density	= 1.70 lb/ft.^{3}

Gas Analysis

Component	Mole Percent
N2	4.0
C1	92.3
C_2	2.4
C_3	0.3
iC_4+	1.0
	100.0

Determine: Design a dry desiccant dehydrator

Solution:

1. Water adsorbed

    For this example, an 8-hour on-stream cycle with 6 hours of regeneration and cooling will be assumed. On this basis, the amount of water to be adsorbed per cycle is:

    $$= \left(\frac{8}{24}\right)(50 \text{ MMSCF})\left(90\frac{\text{lb}}{\text{MMSCF}}\right)$$

    $= 1500 \text{ lb } H_2O/\text{cycle}$

2. Loading

    Because of the relative high operating temperature, use Mobil's SOR beads as the desiccant and design on the basis of 6% loading. SOR beads weigh approximately 49 lbs/ft.^{3} (bulk density). (Refer to Table 2-2.)

    The required weight of desiccant per bed is:

    $$= \frac{1500 \text{ lb } H_2O}{(0.06 \text{ lb } H_2O/\text{lb desiccant})}$$

    $= 25{,}000 \text{ lb desiccant per bed}$

    The required volume of desiccant per bed is:

    $$= \frac{25{,}000 \text{ lb desiccant per bed}}{49 \text{ lb desiccant/ft.}^3}$$

    $= 510 \text{ ft.}^3/\text{per bed}$

3. Tower sizing

    Recommended maximum superficial velocity at 600 psia is about 55 ft./minute (Table 2-1)

    Minimum vessel internal diameter (from Equation 2-1)

$$d^2 = 3600 \left(\frac{Q_g TZ}{VP} \right)$$

$$d^2 = 3600 \frac{(50)(570)(1.0)}{(55)(600)}$$

$d = 55.7$ in or 4.65 ft.

Bed height is

$$L = \frac{50 \text{ ft.}^2}{\frac{\pi(4.65)^2}{4 \text{ ft.}^2}} = 30 \text{ ft.}$$

The pressure drop across a clean bed, assuming ⅛-in bead and $\mu = 0.01$ cp, (Equation 2-2) is

$$\Delta P = |B\mu V = C\rho V^2|L$$

$$\Delta P = \left((0.056)(0.01)(55) + (0.00009)(1.70)(55)^2 \right) 30$$

$$= 14.8 \text{ psi}$$

This is higher than the maximum recommended pressure drop of 8 psi, thus the vessel internal diameter should be increased to the next standard size.

Choose a diameter of 5 ft. 6 inches, substitute into the above equations and determine V, L, and ΔP.

$V = 39.2$ ft./min

$L = 21.5$ ft.

$AP = 5.5$ psi

Allowing 6 ft. space to remove the desiccant and refill would be about 28 ft. This yields an L/D of $28/5.5 = 5.0$ which is acceptable.

4.   Regeneration heat requirement

Assume the bed (and tower) is heated to 350°F. The average temperature will be $(350 + 110)$ °F/2 $= 230$°F.

The approximate weight of 5 ft. 6 in. ID × 28 ft. × 700 psig tower is 53,000 lbs including the shell, heads, nozzles, and supports for the desiccant.

Heating and cooling requirement can be determined using Equation 2-4

$$Q = wC_p\Delta T$$

Where:

Q = Heat required, Btu

C_p = Specific heat of material at average temperature, Btu/lb/°F

Δt = Temperature differential across material, °F

5. Heating requirement/cycle

Desiccant: (25,000 lb)
$(350°F - 100°F)(0.25)^{(4)}$ = 1,500,000 Btu

Tower: (53,000 lb)
$(350°F - 100°F)(0.12)^{(1)}$ = 1,520,000 Btu

Desorb water: (1500 lb)
$(1100 Btu/lb)^{(2)}$ = 1,650,000 Btu

$(1500 lb)(230°F - 110°F)(1.0)^{(3)}$ = 200,000 Btu

Total heat = 4,870,000 Btu

10% heat losses = 490,000 Btu

Total heat requirement/cycle = 5,360,000 Btu/cycle

Notes:

(1) Specific heat of steel.

(2) The number "1100 Btu/lb" is the heat of water desorption, a value supplied by the desiccant manufacturer.

(3) The majority of the water will desorb at the average temperature. This heat requirement represents the sensible heat required to raise the temperature of the water to the desorption temperature.

(4) Specific heat of SOR beads "R" (refer to Table 2-2).

6. Cooling requirement/cycle

Desiccant: (25,000 lb)
$(350°F - 100°F)(0.25)^{(3)}$ = 1,500,000 Btu

Tower: (53,000 lb)
$(350°F - 100°F)(0.12)^{(1)}$ = 1,520,000 Btu

Total cooling = 3,020,000 Btu

10% for non uniform cooling = 300,000 Btu

Total cooling requirement/cycle = 3,320,000 Btu/cycle

This example assumes insulation is on the outside of the towers.

Duty would be less if towers were insulated internally.

Internal insulation should be used to minimize thermal stress caused by wide swings in temperature during regeneration.

Channeling or bypassing gas around the desiccant beds can be a problem.

7.   Regeneration gas heater

Assume inlet temperature of regeneration gas if 400°F.

The initial outlet temperature of the bed will be the bed temperature of 110°F; at the end of the heating cycle, the outlet temperature will be the design value of 350°F. Therefore, the average outlet temperature is (350 + 110) or 230°F.

The volume of gas required for heating will be

$$V_{heating} = \frac{5,360,000 \text{ Btu/cycle}}{(400 - 230)°F(0.64)*\text{Btu/lb/}°F}$$

$$= 49,400 \text{ lbs/cycle}$$

The regeneration gas heater load, Q_H, is then:

$$Q_H = 49,400(400 - 11)(0.62)*\text{Btu/lb/}°F$$

$$= 8,900,000 \text{ Btu/cycle}$$

For design, add 25% for heat losses and nonuniform flow. Assuming a three-hour heating cycle, the regeneration gas heater must be sized for

$$QH = (8,900,000)\left(\frac{1.25}{3}\right)$$

$$= 3,710 \text{ Btu/hr}$$

8.   Regeneration gas cooler

The regeneration gas cooling load is calculated using the assumption that all of the desorbed water is condensed during a half hour of the three-hour cycle. The regeneration gas cooler load Q_c would be:

Regeneration gas: 49,400
$(230 - 110)(0.61)/3$ $= 1,205,000$ Btu/hr

Water: $1500(1157 - 78)^{(1)}/0.5 = 3,237,000$ Btu/hr

Total load $= 4,442,000$ Btu/hr

10% heat loss $= 44,000$ Btu/hr

Total $= 4,886,000$ Btu/hr

9.   Cooling cycle

Similarly for the cooling cycle where the initial outlet temperature is 350°F and at the end of the cooling cycle, it is approximately 110°F.

The average outlet temperature is $(350 + 110)/2 = 230$°F.

Assuming the cooling gas is at 110°F, the volume of gas required for cooling will be

$$V_{cooling} = \frac{3,320,000 \text{ Btu/cycle}}{(230 - 100)°F(0.59)*\text{Btu}/°F}$$

$$= 46,900 \text{ lbs/cycle}$$

Notes: (1) from steam tables

ABSORPTION

Process Overview

In the absorption process, a hygroscopic liquid is used to contact wet gas and remove the water vapor.

The most common liquid used in absorption type dehydration units is triethylene glycol (TEG).

Important physical properties of glycols are shown in Part 1.

Principles of Absorption

Absorption and Stripping

Through absorption, the water in a gas stream is dissolved in a relatively pure liquid solvent stream.

The reverse process, in which the water in the solvent is transferred into the gas phase, is known as stripping.

The terms regeneration, reconcentration, and reclaiming are also used to describe stripping (or purification) because the solvent is recovered for reuse in the absorption step.

Absorption and stripping are frequently used in:

Gas processing

Gas sweetening

Glycol dehydration

Raoult and Dalton's Laws

Absorption can be qualitatively modeled by using Raoult's and Dalton's laws.

For a vapor liquid equilibrium system:

Raoult's Law state that the partial pressure of a component in a vapor phase that is in equilibrium with a liquid is directly proportional to the mole fraction of the component in the liquid phase.

Dalton's Law states that the partial vapor pressure of a component is equal to the total pressure multiplied by its mole fraction in the gas mixture.

Raoult's Law expressed in equation form is:

$$p_i = P_i X_i \tag{2-5}$$

Dalton's Law expressed in equation form is:

$$p_i = P Y_i \tag{2-6}$$

Where:

p_i = Partial vapor pressure of component i

P_i = Vapor pressure of pure component i

X_i = Mole fraction of component i in the liquid

P = Total pressure of the gas mixture

Y_i = Mole fraction of component i in the vapor

Combining these laws we have:

$$PY_i = p_i X_i$$

or

$$p_i/P = Y_i/X_i \tag{2-7}$$

Since the pure-component vapor pressure and the total pressure are not affected by composition. Equation (2-7) is significant.

It states that the ratio of the vapor mole fraction to the liquid mole fraction for any component is independent of the concentrations of that component and the other components present.

The ratio Y_i/X_i is commonly known as the K-value.

Since the pure component vapor pressure increases with temperature, the K-value increases with increasing temperature and decreases with increasing pressure. In physical terms this means:

> Transfer from the gas phase to the liquid phase (absorption) is more favorable at lower temperature and high pressures

> Transfer to the gas phase (stripping) is more favorable at higher temperatures and lower pressures

Glycol-Water Equilibrium

Absorption processes are dynamic and continuous.

> Gas flow cannot be stopped to let the vapor and liquid reach equilibrium.

> Thus, the system must be designed to approach equilibrium as closely as possible while flow continues.

> This is accomplished by using a trayed or packed contactor in which the gas and liquid are in counter current flow.

The closer to 100% equilibrium that a tray or packed section approaches, the higher the tray or packing efficiency. For example,

> A common tray efficiency is 25%, meaning that 25% of the water molecules that would have been transferred under equilibrium conditions were actually transferred.

> Wet gas enters the bottom of the column and contacts the rich glycol (high water content) just before the glycol leaves the column.

> The gas encounters leaner glycol as it works its way up the column, contacting the leanest glycol (lowest water content) just before it leaves the column.

The equilibrium based on Dalton's and Raoult's Laws can be rearranged as follows:

$$Y_i = X_i \left(\frac{P_i}{P} \right) \tag{2-8}$$

Since P_i/P is constant for constant temperature, the concentration of the water in the gas must be directly proportional to the concentration in the liquid.

However, the liquid concentration is constantly changing as water is absorbed.

The counter current flow in the contactor makes it possible for the gas to transfer a significant amount of water to the glycol and still approach equilibrium with the leanest glycol concentration.

GLYCOL DEHYDRATION

Principles of Operation

Introduction

After the liquid (free) water has been removed from the gas stream by separation, 25 to 120 lbs of water per MMSCF of gas will remain, depending on the temperature and pressure of the gas.

The warmer the inlet gas and the lower the pressure, the more water vapor the gas stream will contain (see Figure 2-13).

Normally, between 20 to 115 lbs of water per MMSCF of gas must be removed before the required dew point of the gas is met.

The schematics in Figures 2-14 and 2-15 show the flow through a typical glycol dehydration system. The glycol dehydration process can be discussed in two parts:

Gas system (Figure 2-14)

Glycol system (Figure 2-15)

Gas System

Inlet Scrubber/Microfiber Filter Separator

Wet gas enters the unit through the inlet gas scrubber/microfiber filter separator, usually vertical, to remove liquid and solid impurities.

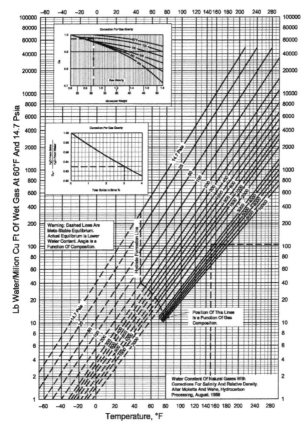

FIGURE 2-13 Water content of sweet, lean natural gas – McKetta-Wehe.

Glycol Gas Contactor

After passing through the microfiber filter separator, the gas enters the glycol gas contactor near the bottom of the vessel.

The inside of the contactor contains either packing of several trays with weirs that maintain a specific level of glycol so that the gas must bubble through the glycol as the gas flows up (Figure 2-16).

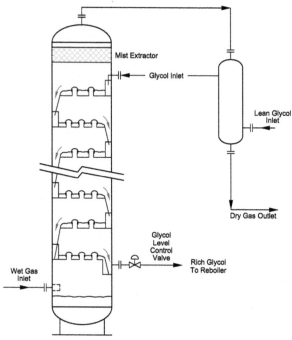

FIGURE 2-14 Gas system.

As the wet gas passes upward through each succeeding tray, it gives up the water vapor to the glycol and becomes progressively drier.

Before leaving the contactor the gas passes through a mist extractor to remove glycol that may be trying to leave the gas.

Dry gas exits the contactor at the top and passes through an external glycol gas heat exchanger where it cools the incoming dry glycol to increase its absorption capacity (Figure 2-17).

Some installations incorporate a glycol knockout drum (centrifugal separator) which recovers any glycol that has escaped with the gas through the mist extractor (Figure 2-18).

The dry gas then leaves the dehydrator unit.

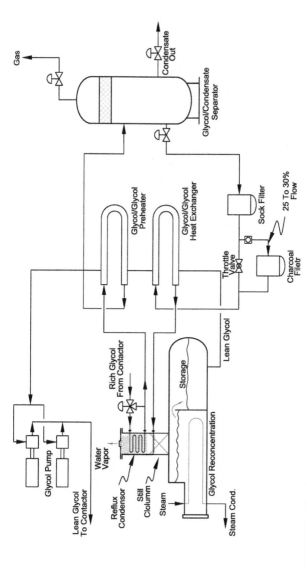

FIGURE 2-15 Glycol system.

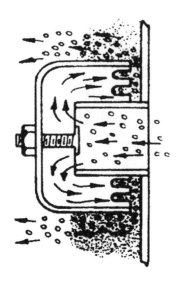

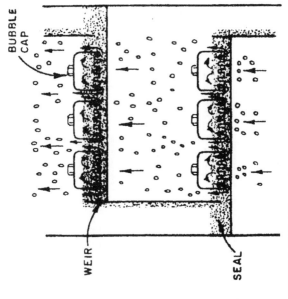

FIGURE 2-16 Bubble cap trays.

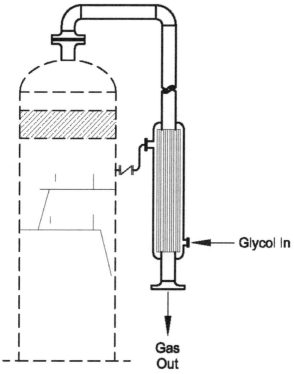

FIGURE 2-17 External glycol gas heat exchanger.

Glycol System

Glycol Gas Heat Exchanger

Dry concentrated glycol is pumped up to contactor pressure, by the glycol pump, and then passes through the glycol gas heat exchanger before entering the contractor tower.

The glycol gas heat exchanger cools the glycol to near the temperature of the gas before the glycol enters the contactor.

It is important that the glycol be near the gas temperature to:

Prevent gas from exceeding equilibrium temperature

Prevent foaming

FIGURE 2-18 Centrifugal separator used to recover escaping glycol with gas.

Glycol Gas Contactor

Dry glycol from the glycol gas heat exchanger enters the contactor tower and flows across the top tray.

This is the first contact between the glycol and gas.

Glycol flows downward through downcomers in the tower, absorbing more water as it passes across each tray.

The downcomer seals the glycol passage into the tray below, thus preventing gas from short-circuiting past the bubble caps.

As the glycol flows downward through each succeeding tray, it becomes wetter with the water it has absorbed from the gas and collects in the bottom of the contactor saturated with water.

As the gas moves upward through each succeeding tray, it becomes drier.

The wet gas that has accumulated in the bottom of the contactor passes through a strainer (filter), which removes abrasive particles, before flowing through the power side of the glycol pump (energy exchange pumps), where it furnishes the power to pump the dry glycol into the contactor.

Power comes from the increased head caused by the absorbed gas contained in the rich glycol.

Reflux Condenser

From the glycol gas contactor the cool wet glycol passes through a coil (reflux condenser) in the top of the reboiler still column.

The coil cools the vapors leaving the still column and condenses the glycol vapors to liquid.

The glycol liquid droplets gravitate back down the still column to the reconcentrator.

The water remains as a vapor and continues on out the top of the still column.

The cooling coil is commonly called the reflux condenser.

Glycol-Glycol Preheater

The slightly warmed wet glycol leaving the reflux condenser passes through the glycol-glycol preheater.

The hot dry glycol from the glycol reconcentrator heats the wet glycol further, and in turn further cools the dry glycol before it goes to the glycol pumps.

Gas-Glycol-Condensate Separator

After leaving the glycol-glycol preheater, the heated wet glycol is sent to a low-pressure gas-glycol-condensate separator, where most of the entrained gas and liquid hydrocarbons that were picked up by the glycol on its path through the contactor are removed.

The heat provided by the glycol-glycol preheater helps in the separation of hydrocarbons from the wet glycol.

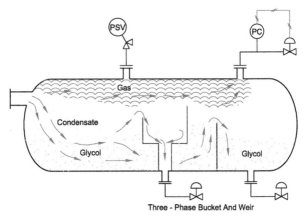

FIGURE 2-19 Gas-glycol-condensate separator.

The hydrocarbon condensate is separated from the glycol by a three-phase gas-glycol-condensate separator (Figure 2-19).

Microfiber Filter

After the gas and condensate has been separated in the gas-glycol-condensate separator, the wet glycol passes through a microfiber filter (Figure 2-20).

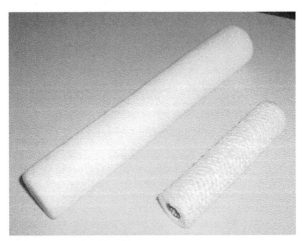

FIGURE 2-20 Microfiber filters.

These filters are used to remove solids, tarry hydrocarbons, or other impurities.

Charcoal (Carbon) Filter

From the microfiber filter the wet glycol enters a charcoal or carbon filter.

Activated carbon granules in this filter absorb liquid-entrained hydrocarbons, well-treating chemicals, compressor oils, and other impurities that may cause foaming.

Glycol-Glycol Heat Exchanger

From the charcoal filter, the wet glycol flows through the dry glycol to the wet glycol heat exchanger.

This heat exchanger preheats the wet glycol as much as possible before entering the glycol reconcentrator, thus reducing the heat duty of the glycol reconcentrator.

Still Column

From the glycol/glycol heat exchanger, the wet glycol enters the still column which sits vertically atop the glycol reconcentrator (Figure 2-21).

The inside of the still column is packed with either ceramic saddles or stainless steel pall rings, which are used to add surface area and distribute heat to the incoming glycol.

The incoming wet glycol spreads out uniformly and drips down through the packed section.

The vapors traveling upward from the glycol reconcentrator heats the packing.

As the glycol travels down through the heated packing, water begins to be driven off as steam.

Units utilizing efficient heat exchangers may remove as much as 75 to 80% of the water contained in the glycol in the still column before the glycol reaches the reconcentrator.

As water vapor travels up through the still column and exits from the top, it carries with it trapped glycol vapor.

To prevent the loss of glycol vapor, the still column utilizes a "reflux condenser" located on the top of the packed still column.

Glycol vapors escaping the still column with the steam are attracted to the film of condensed liquid

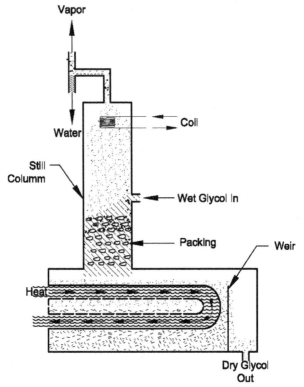

FIGURE 2-21 Still column at top of glycol reconcentrator.

(primarily water) covering the coil surface area where they too are condensed.

The liquid droplets gravitate back down the still column into the reconcentrator for further treating, thus preventing excessive glycol loss due to vaporization.

On some units, the glycol enters the still column below the packed section of the column.

Vaporization takes place in the reconcentrator.

The reflux condenser operates the same in both types of still columns.

The packed section is no longer used to distribute heat for vaporization.

Condensed liquid from the reflux condenser drops back into the packed section providing a liquid film over the upper portion of the packing.

Glycol vapors escaping with steam from the reconcentrator must pass through the packed section.

The watery film covering the packing recaptures the glycol vapor, condensing it into droplets, which wash back into the reconcentrator.

Thus, more glycol vapor can be recovered in this configuration than in the previously described still column.

Since vaporization occurs primarily in the reconcentrator, the operating temperature is lower in this type of still column. This translates into:

Greater reflux condensation

Requires larger heat duty

Reconcentrator

From the packed still column, the wet glycol drops downward into the reconcentrator.

The glycol is heated to a temperature at which most of the remaining water and some of the glycol are vaporized.

A heat source heats the glycol to between 350° and 400°F.

It removes the remaining water

It is below the decomposition point of TEG.

The temperature of the glycol in the reconcentrator is critical and must be controlled at this point.

Sources of heat include:

Direct fired (natural draft/forced draft)

Waste heat (exhaust gases from compressors or generators)

Electric heaters

The heated vapor (both glycol and water) rises upward through the still column.

As the mixture passes the cool reflux condenser coils, the glycol vapors are condensed and drop back down.

The water vapor leaves the top of the still column as steam.

Some of the steam will condense, so a downspout is provided to drain the water off.

A weir maintains a level of glycol over the heat source, which:

> Prevents over heating of the tubes

> Prevents premature tube failure

As the glycol is purified, it spills over the weir into a separate compartment.

From the reconcentrator, the dry (lean) glycol flows to the accumulator surge tank, when the glycol pump raises it to contactor pressure to start another cycle.

Stripping Gas

Purities of 98% or more are normally achieved in a TEG system operating at atmospheric pressure.

If very pure glycol (up to 99.9% TEG) is required and cannot be achieved by the standard regeneration system, stripping gas may be used.

A small amount of dry natural gas, normally taken from the fuel stream, is injected into the reconcentrator.

Since hot gas has an affinity for water, the stripping gas is bubbled through the hot glycol, which strips the remaining water from the glycol.

This gas can be put directly into the reconcentrator or it can be added to the storage tank where it can percolate through the packed column between the two vessels (Stahl column).

The Stahl column also serves as a weir where the dry glycol spills downward by gravity over packing while the gas goes upward, removing even more water.

This method prevents air from coming into contact with the dry glycol in the storage tank, thus preventing oxidization of the glycol.

Oxygen entry into the glycol system will:

> Decompose the glycol to some extent

> Cause corrosion within the system.

Stripping gas can:

> Reduce the temperature at which the reconcentrator must operate

> Reduce the glycol circulation rate necessary to dehydrate the gas adequately

Effect of Operating Variables

General Considerations

Several operating and design variables have an important effect on the successful operation of a glycol dehydration system.

Glycol Selection

Glycols are the most commonly used liquid desiccants in the absorption process because they are:

> Highly hygroscopic (readily absorb and retain water)

> Stable to heat and chemical decomposition at the temperature and pressures necessary in the process

> Low vapor pressures, which minimize equilibrium loss of the glycol in the residual natural gas stream and in the regeneration system

> Easily regenerated (water removed) for reuse

> Noncorrosive and nonfoaming at normal conditions; impurities in the gas stream can change this, but even then inhibitors can help to minimize these problems

> Readily available at moderate cost

Hygroscopicity of glycols is affected by the concentration (glycol-to-water ratio), that is, increasing as the concentration increases.

Dew point depression obtainable in a gas stream increases as the glycol concentration increases.

Ethylene Glycol (EG)

Ethylene Glycol tends to have high vapor losses to gas when used in a contactor.

It is used as a hydrate inhibitor where it can be recovered from the gas by separation at temperatures below 50°F.

Diethylene Glycol (DEG)

Diethylene Glycol reconcentrates at temperatures between 315° and 325°F, which yields purity of 97.0%.

It degrades at 328°F.

It cannot achieve the concentration required for most applications.

Triethylene Glycol (TEG)

Triethylene Glycol is most commonly used in glycol dehydration.

It reconcentrates at temperatures between 350° and 400°F, which yields purity of 98.8%.

It degrades at 404°F.

It tends to experience high vapor losses to gas at temperatures in excess of 120°F.

With stripping gas, dew point depressions up to 150°F are possible.

Tetraethylene Glycol (TTEG)

Tetraethylene Glycol is expensive.

It reconcentrates at temperatures between 400° and 430°F.

It experiences lower vapor losses to gas at high gas contactor temperatures.

It degrades at 460°F.

Inlet Gas Temperature

At constant pressure, the water content of the inlet gas increases as the temperature increases. For example, at 1000 psia and

80°F, gas holds 34 lb of water/MMSCF

120°F, gas holds 104 lb of water/MMSCF

If the gas is saturated at the higher temperature, the glycol will have to remove about three times as much water to meet the specifications.

Temperatures above 115°F result in high glycol losses, thus requires tetraethylene glycol.

Temperature should not fall below the hydrate formation temperature range (65° to 70°F) and always above 50°F.

Temperatures below 50°F cause problems due to an increase in glycol viscosity.

Temperatures below 60° to 70°F can cause a stable emulsion with liquid hydrocarbons in the gas and cause foaming in the contactor.

An increase in gas temperature increases the gas volume, which in turn increases the diameter of the glycol contactor.

Lean Glycol Temperature

Dry glycol temperatures entering the top tray of the contactor (approach temperature) should be held low (10° to 15°F) above the inlet gas temperature.

Equilibrium conditions between the glycol and the water vapor in the gas is affected by temperature.

Glycol entering the top tray of the contactor may raise the temperature of the gas surrounding it and prevent the gas giving up its remaining water vapor.

Inlet glycol temperatures greater than 15°F above the gas temperature results in high glycol losses to the gas.

Drastic temperature differential also has a tendency to emulsify the glycol with any contaminants subsequent glycol loss.

Glycol Reconcentrator Temperature

Reconcentrator temperature controls the concentration of the water in the glycol.

With a constant pressure, the glycol concentration increases with higher reconcentrator temperature.

Reconcentrator temperature should be limited to between 350° and 400°F.

Minimizes degradation of TEG which begins to degrade at 404°F

Results in lean glycol concentrations between 98.5 and 98.9%

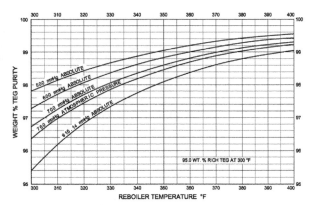

FIGURE 2-22 Glycol purity versus reconcentrator temperature at different levels of vacuum.

Figure 2-22 shows the glycol concentrations that can be obtained with various reboiler temperatures.

When higher lean glycol concentrations are required:

Add stripping gas to the reconcentrator, or

Operate the reconcentrator and still column in a vacuum.

Temperature at Top of Still Column

A high temperature in the top of the still column can increase glycol losses due to excessive vaporization.

A reboiler temperature in the range of 350° to 400°F insures adequate heat transfer to the ceramic packing in the still column.

The still column operates best (allows the steam to escape) when the vapor outlet temperature is between 215° and 225°F.

When the temperature reaches 250°F and above, glycol vaporization losses increase.

Still top temperature can be lowered by increasing the amount of glycol flowing through the reflux condenser coil.

If the temperature in the top of the still column drops too low, (below 220°F) too much water can be condensed and washed back into the reconcentrator, which increases the reconcentrator heat duty.

Too much cool glycol circulation in the reflux condenser coil can lower the still top temperature below 220°F which can cause the excess water to condense.

> Thus, most reflux condenser coils have a bypass valve, which allows manual or automatic control of the stripping still temperature.

Contactor Pressure

At a constant temperature, the water content of the inlet gas decreases with an increase in pressure.

The lower the pressure, the larger the contactor diameter required.

Good dehydration can be achieved at any pressure below 3000 psig as long as the pressure is constant.

Optimum dehydration pressure is often in the range of 550 to 1200 psig.

Sizing calculations should always be based on minimum expected operating gas pressure.

Rapid pressure changes translate into rapid velocity changes in the contactor which:

> Breaks the liquid seals between the downcomers and the trays

> Allows the gas to channel up through both the downcomer and bubble caps

> Allows the glycol to be swept out with the gas

Reconcentrator Pressure

Reducing the pressure in the reconcentrator at a constant temperature results in higher glycol purity.

Most reconcentrators operate between 4 to 12 ounces of pressure.

On standard atmospheric reconcentrators, pressures in excess of 1 psi results in:

> Glycol loss from the still column

> Reduction of lean glycol concentration

> Reduction in dehydration efficiency

Pressures more than 1 psi are usually associated with excess water in the glycol and create a vapor velocity exiting the still great enough to sweep glycol out.

Fouled still column packing often contributes to high reconcentrator pressure.

Still column should be adequately vented and packing replaced periodically so as to prevent back pressure on the reconcentrator.

Pressures below atmospheric will increase the lean glycol concentration because the boiling temperature of the rich glycol/water mixture decreases.

Reconcentrators are rarely operated in a vacuum due to the added complexity and the fact that air leaks will result in glycol degradation.

Contractor Pressure

If lean glycol concentrations in the range of 99.5% are required consider:

Operating the reconcentrator at a pressure 500 mm Hg absolute (10 psia), or

Using stripping gas

Figure 2-22 can be used to estimate the effect of operating in a vacuum on lean glycol concentration.

Glycol Concentration

The water content of the dehydrated gas depends primarily on the lean glycol concentration.

The higher the concentration of lean glycol entering the contactor, the greater the dew point depression for a given circulation rate and number of trays.

Increasing the glycol concentration above a 99% purity can lead to dramatic results on the outlet dew point (Figure 2-23). For example, with a 100°F inlet gas temperature (110°F top tray temperature), an outlet dew point of

10°F can be obtained with 99.0% TEG

−30°F can be obtained with 99.8% TEG

−40°F can be obtained with 99.9% TEG

Higher concentrations of TEG can be obtained by:

Increasing the glycol reconcentration temperature

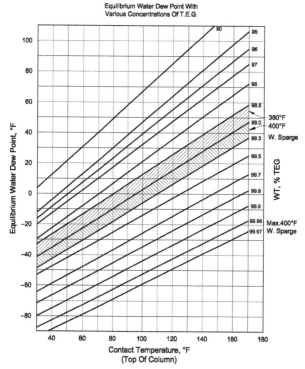

FIGURE 2-23 Equilibrium water dew points with various concentrations of TEG.

Injecting stripping gas into the reconcentrator

Reducing the operating pressure of the reconcentrator

Reconcentration temperatures for TEG normally run between 380°F and 400°F, which results in glycol purities of 98% to 99%. Figures 2-24 and 2-25 illustrate the effect of stripping gas.

If gas is injected directly into the reconcentrator (via a sparger tube), the concentration of TEG increases significantly from 99.1% to near 99.6% as the gas rate is increased from 0 to 4 SCF/gal.

When the Stahl method is used (counter current gas stripping after the reconcentrator),

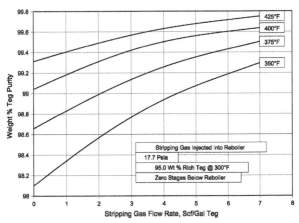

FIGURE 2-24 Effect of stripping gas on TEG concentration.

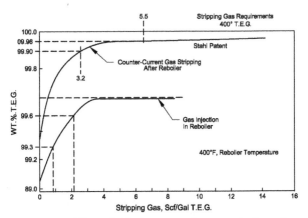

FIGURE 2-25 Effect of stripping gas on the concentration using Stahl column.

concentrations as high as 99.95% TEG can be attained at a 400°F reconcentrator temperature.

Glycol Circulation Rate

When the number of absorber trays and lean glycol concentration are held constant, the dew point depression of a saturated gas is increased as the glycol circulation rate is increased.

The more lean glycol that comes into contact with the gas, the more water vapor is absorbed out of the gas.

Whereas the glycol concentration mainly affects the dew point of the dry gas, the glycol rate controls the total amount of water that can be removed.

The normal operating level in a standard dehydrator is 3 gallons of glycol per pound of water removed (Range 2-7).

Figure 2-26 shows that a greater dew point depression is easier to achieve by increasing the glycol concentration rate.

Excessive circulation rates:

> Overload the reconcentrator
>
> Prevent good glycol regeneration
>
> Prevent an adequate glycol gas contact in the absorber
>
> Increase pump maintenance problems
>
> Increase glycol loses

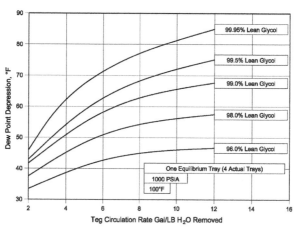

FIGURE 2-26 Calculated dew point depression versus circulation rate (1 equilibrium tray (4 actual trays)).

Heat required by the reboiler is directly proportional to the circulation rate. An increase in circulation rate may:

Decrease the reconcentrator temperature

Decrease the lean glycol concentration

Decrease the amount of water that is removed by the glycol from the gas

Only if the reconcentrator temperature remains constant will an increase in circulation rate lower the dew point of the gas.

Number of Absorber Trays

When the glycol circulation rate and the lean glycol concentration are held constant, the dew point depression of a saturated gas is increased as the number of trays is increased.

Actual trays do not reach equilibrium, and their approach to it is expressed as a fraction of a theoretical tray.

A tray efficiency of 25% is commonly used for design.

Four actual trays with efficiencies of 25% would accomplish the job of one theoretical tray.

The number of actual trays in a design ranges from 4 to 12.

An approximation of the actual number of valve trays per foot of packing can be obtained from Figure 2-27.

For high performance units, the specification of more than 4 trays in a new design can achieve fuel savings (for the same dew point depression) due to

Lower circulate rate

Lower reconcentration temperature

Lower stripping gas rate

Figure 2-28 shows that specifying a few additional trays in the contactor is much more effective than increasing the glycol circulation rate. The additional investment for a taller contactor is often justified by fuel savings.

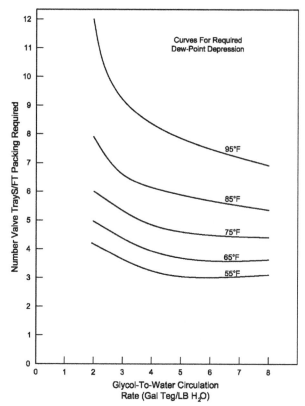

FIGURE 2-27 Trays of packing required for glycol dehydrator.

SYSTEM DESIGN

Sizing Considerations

Involves specifying the following:

Glycol gas contactor diameter

Number of absorber trays (which establishes the tower's overall height)

Glycol circulation rate

Lean glycol concentration

Reconcentrator heat duty

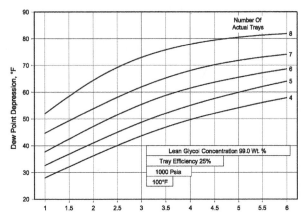

FIGURE 2-28 Effect of number of absorber trays on dew point depression.

The number of absorber trays, glycol circulation rate, and lean glycol concentration are all inter-related.

Inlet Microfiber Filter Separator

Clean inlet gas is a key factor in minimizing absorber operating problems.

By eliminating liquid water carry-over, the inlet scrubber (microfiber filter separator) can prevent the following problems:

> Dilution of the glycol
>
> Lower absorber efficiencies
>
> High glycol circulation rates
>
> High vapor-liquid loads on the still column
>
> Still column flooding
>
> High reboiler heat-load and fuel-gas requirements

These problems also cause:

> High glycol losses
>
> Wet sales gases

The scrubber also prevents salt or other solids from entering the glycol system, where they could be deposited in the reconcentrator to:

Foul the heating surfaces

Burn out as hot spots

It should be sized according to procedures for gas-liquid separation sizing.

The vessel cleans the incoming gas and is useful when paraffin and other impurities that are present in a fine vapor form:

A mist extractor sized to remove 99% of all contaminants over one micron in size

A microfiber filter separator is highly recommended when inlet gas streams have been compressed or if structured packing is used in the contractor.

Compressor oil and heavy distillate can coat the tower packing either in the contactor or still column, thus decreasing the effectiveness of the equipment.

Glycol Gas Contactor

There are two basic types of glycol gas contactor:

Trayed towers

Packed towers

Triethylene glycol is:

Viscous (resulting in poor tray efficiency)

Exhibits foaming tendencies (restricts tower performance)

Due to the low liquid loadings, it is usually possible to provide a high downcomer residence time with only a small percentage of tower area dedicated to downcomers.

Some contactors have an "internal scrubber" which occupies approximately the lower one-third of the vessel.

They are usually installed on units where the inlet gas flow rate is less than 50 MMSCFD.

"Chimney" is included on the scrubber/contactor combination (Figure 2-29):

Consists of a large stack that covers the top of the inlet scrubber

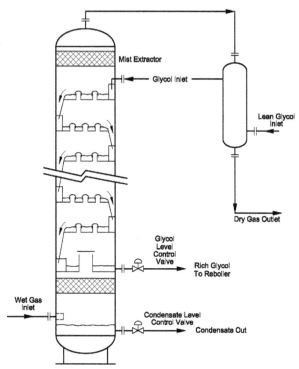

FIGURE 2-29 Glycol contactor with scrubber section.

Allows the gas to pass upward from the scrubber section to the absorber section

Prevents glycol from being lost out of the scrubber section

Some contact towers have an internal three phase separator:

Distinguishable in that the lower section has two sets of level controls and two liquid dump valves

Not recommended as it is difficult to troubleshoot should a problem occur

A separate two-phase microfiber filter separator located immediately upstream of the contactor is the most efficient configuration

Contactor Diameter

The minimum diameter for trayed towers and conventional packing can be determined from the following equation:

$$d^2 = 5040\left(\frac{T_0 Z Q_g}{P}\right)\left|\left(\frac{\rho_s}{\rho_L - \rho_g}\right)\left(\frac{C_D}{d_m}\right)\right|^{1/2} \qquad (2\text{-}9)$$

Where:

d = Contactor inside diameter, inches

d_m = Drop size, microns = 120 to 150 micron range

T_o = Contactor operating temperature, °R

Q_g = Design gas flow rate, MMSCFD

P = Contactor operating pressure, psia

C_d = Drag coefficient

ρ_g = Gas density, lb/ft.^{3} = 2.7 (SP/TZ)

ρ_L = Glycol density, lb/ft.^{3} = 70 lb/ft.^{3}

Z = Compressibility factor

S = Gas specific gravity (air = 1)

Structured packing can handle higher gas flow rates for the same diameter contactor. Figures 2-30 through 2-33 are correlations prepared by vessel manufacturers that allow graphical solutions of glycol gas contactor diameters.

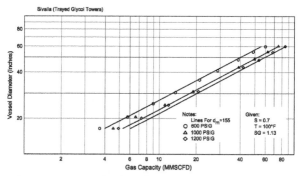

FIGURE 2-30 Determination of contactor diameter—Sivills.

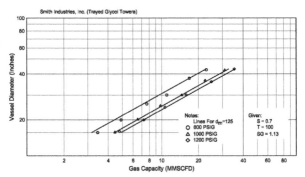

FIGURE 2-31 Determination of contactor diameter—Smith Industries.

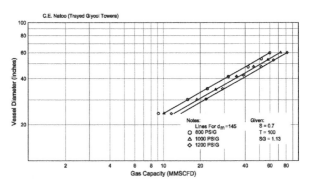

FIGURE 2-32 Determination of contactor diameter—NATCO.

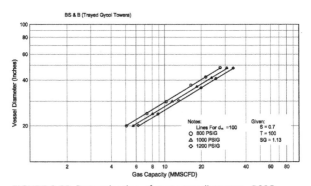

FIGURE 2-33 Determination of contactor diameter—BS&B.

Tray Design

Bubble Cap Trays (Figures 2-34 through 2-38)
Most commonly used design

Better than conventional packing (Figure 2-39)

FIGURE 2-34 Common bubble cap configurations.

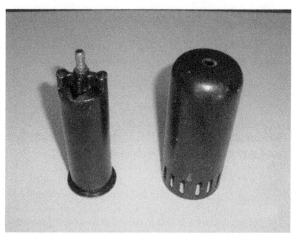

FIGURE 2-35 Bubble cap components.

FIGURE 2-36 Bubble cap tray outside the contactor tower.

FIGURE 2-37 Bubble cap tray inside the contactor tower.

Valve or Flapper Trays (Figures 2-40 through 2-42)
Gas moves upward through a hole in the bottom of the tray.

Over the hole is a device that flutters or flaps in an "up and down" manner breaking the gas stream into bubbles which form the froth layer.

FIGURE 2-38 Bottom of bubble cap tray.

Perforated (Sieve) Trays (Figure 2-43)
> Tray consisting of hundreds of tiny holes
>
> Gas stream that passes through these holes breaks up into the bubbles necessary to form a froth
>
> Inexpensive to fabricate
>
> Gas capacity range that can be effectively dehydrated is limited

Structured (Matrix) Packing (Figure 2-44 and 2-45)
> The matrix resembles corrugated metal affixed side-by-side with the corrugations set at opposite angles.
>
> Gas passes upward, through small holes that are drilled in the corrugations, and forms channels by the opposing corrugation.
>
>> Glycol runs down through the holes and channels contacting the gas.
>
> This is the most efficient packing.

Tray Spacing

> Range 20 to 30 inches
>
> 24 inches is preferred.

Lessig Ring

Raschig Ring

Splined Ring

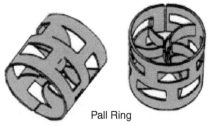

Pall Ring

FIGURE 2-39 Various types of conventional packing.

30-inch spacing is recommended if foaming is
anticipated

Number of Trays

6 to 8 trays are used to meet normal dew-point
depressions.

12 trays are typically required for high dew-point
depressions.

Downcomers

Sized for a maximum velocity of 0.25 ft./sec.

FIGURE 2-40 Top of flapper tray.

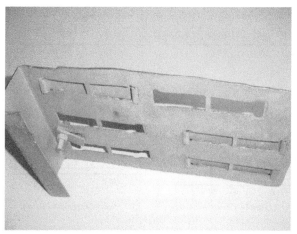

FIGURE 2-41 Bottom of flapper tray.

Glycol Circulation Rate

For a given dew-point depression, the circulation rate is dependent upon:

Lean glycol concentration

Number of trays

FIGURE 2-42 Top and bottom of valve tray.

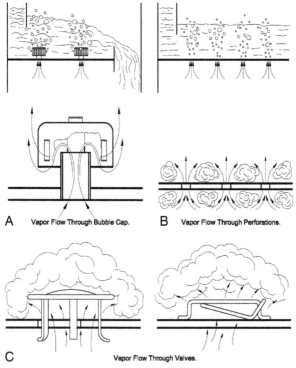

A   Vapor Flow Through Bubble Cap.

B   Vapor Flow Through Perforations.

C   Vapor Flow Through Valves.

FIGURE 2-43 Types of valve trays.

FIGURE 2-44 Structured packing—side view.

FIGURE 2-45 Structured packing—top view.

When the lean glycol concentration and number of trays are held constant, the required glycol circulation rate can be determined from the following equation:

$$L = \frac{\left(\dfrac{\Delta W}{W_i}\right) W_i Q_g}{24} \qquad (2\text{-}10)$$

Where:

$$L = \text{Glycol circulation rates, gal/hr}$$

$$\frac{\Delta W}{W_i} = \text{Circulation ratio, gal TEG/lb } H_2O \text{ (see Figures 2-46, 2-47, and 2-48)}$$

$$W_i = \text{Water content of inlet gas, lb } H_2O/\text{MMSCF}$$

$$W_0 = \text{Desired outlet water content, lb } H_2O/\text{MMSCF}$$

$$\Delta W = W_i - W_0$$

$$Q_g = \text{Gas flow rate, MMSCFD}$$

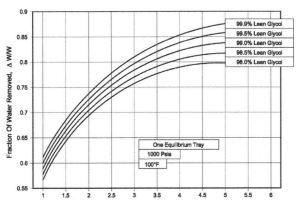

FIGURE 2-46 Fraction of water removed versus TEG circulation rate ($n = 1$ theoretical trays, 4 actual trays).

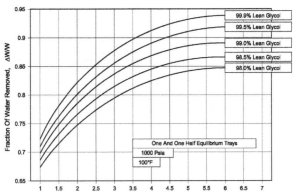

FIGURE 2-47 Fraction of water removed versus TEG circulation rate ($n = 1\frac{1}{2}$ theoretical trays, 6 actual trays).

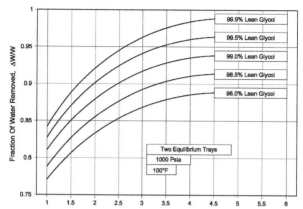

FIGURE 2-48 Fraction of water removed versus TEG circulation rate ($n = 2$ theoretical trays, 8 actual trays).

Figures 2-46 through 2-48 show the fraction of water removed versus TEG rate with respect to different glycol purities.

Lean Glycol Concentration

Equilibrium water dew points for various concentrations of TEG are shown in Figure 2-23.

Glycol purity (lean glycol concentration) is a function of the temperature of the reconcentrator (Figure 2-22).

Glycol purity can be increased by:

Adding stripping gas

Reducing the pressure in the reconcentrator

Reducing the glycol circulation rate

Glycol-Glycol Preheater

Cool wet glycol from the contactor enters the preheater (heat exchanger) at 100°F and the warm glycol leaves at 175° to 200°F en route to the gas/glycol/condensate separator.

Hot dry glycol from the glycol/glycol heat exchanger enters the preheater at 250°F and the warm dry glycol leaves at 150°F to the glycol pumps en route to the contactor.

Temperature limitations to the glycol pump:

Glycol powered pumps (Kimray) limited to 200°F.

Electric plunger pumps limited to 250°F.

Overall heat transfer coefficient (U = 10 to 12)

Glycol-Gas Cooler

TEG to gas contactor is limited to 10°F to 15°F above the inlet gas temperature.

If hotter, some TEG will vaporize with gas.

If colder, gas condensation of the hydrocarbons may cause foam and glycol loss.

Overall heat transfer coefficient (U = 45).

Glycol-Glycol Heat Exchanger

Hot dry glycol from the reconcentrator enters the heat exchanger at 390°F and leaves at 250°F enroute to the glycol/glycol preheater.

Warm wet glycol from the charcoal filter enters the heat exchanger at 200 °F and the hot wet glycol leaves at 350°F en route to the still column.

Gas-Glycol-Condensate Separator

Separator should be sized using procedures for sizing gas-liquid separation.

Liquid retention times between 20 and 30 minutes, depending on API gravity of the condensate, are recommended.

Operating pressure of 35 to 50 psig is recommended.

Reconcentrator

The reconcentrator should be designed to operate

350° to 400°F with TEG, and

305°F with DEG

Design temperature should be sufficiently below the decomposition point so that hot spots on the fire tube and poor mixing in the reconcentrator will not cause decomposition of the glycol.

With everything else operating normally, the reconcentrator temperature is raised to lower the water content of the treated gas, and vice versa.

Specific reconcentrator operating temperature is determined by trial and error.

Temperatures up to 400°F are common

400°F yields 99.5% TEG purity

375°F yields 98.3% TEG purity

Heat Duty

Estimated from the following equations

$$q_t = LQ_L \tag{2-11}$$

Where:

q_t = Total heat duty on reconcentrator, Btu/hr

L = Glycol circulation rate, gal/hr

Q_L = Reconcentrator heat load, Btu/gal TEG

Table 2-4

Heat duty estimated from Equation (2-11) is normally increased by 10 to 25% to account for start-up, fouling, and increased circulation.

Fire Tube Sizing

The actual surface area of the firebox required for direct-fired heaters can be calculated from the following equation:

Table 2-4 Reconcentrator Heat Load

Design (gal TEG/lb H_2O)	Reconcentrator Heat Load (Btu/gal TEG circulated)
2.0	1066
2.5	943
3.0	862
3.5	805
4.0	762
4.5	729
5.0	701
5.5	680
6.0	659

$$A = \frac{q_t}{6000} \tag{2-12}$$

Where:

A = Total firebox surface area, ft.^{2}

q_t = Total heat duty on reconcentrator, Btu/hr

By determining the diameter and overall length of the U-tube fire tube, one can estimate the overall size of the reconcentrator.

A heat flux of 6000 to 8000 Btu/hr-ft.^{2} is often used, but the 6000 value is suggested to ensure against glycol decomposition.

Reflux Condenser

Wet glycol inlet from the gas contactor enters at 115°F and leaves at 125°F.

Controls TEG losses

Reflux rate should be 50% of the water removal rate.

Condensing coil:

Allows uniform year round refluxing

Provides lowest TEG loss

Provides most economical reconcentrator operation

Stripping Still Column

Temperature considerations:

Temperature is critical to the operation of the still column.

Heat is provided by the reconcentration.

Reconcentrator temperatures in the range of 350°F to 400°F insures adequate heat transfer to the ceramic packing in the still column.

Still columns whose wet glycol inlet enters above the packed section (Figure 2-49):

Operate best with a vapor outlet temperature between 225° and 250°F

Purpose of glycol falling over ceramic packing is the efficient use of available heat

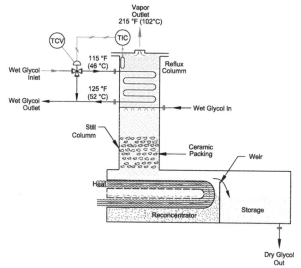

FIGURE 2-49 Still column with wet glycol entering above the ceramic saddle packing.

Backpressure should be kept to a minimum (1 psig is maximum)

Still columns whose wet glycol inlet enters below the packed section (Figure 2-50):

Allow pall ring type packing to be solely involved in the reflux process

Operate best with a vapor outlet temperature between 185°F and 195°F

This temperature allows a greater volume of condensation by the reflux coil while still permitting the majority of the steam to escape

Diameter Size

Diameter size is based on the required diameter at the base of the still, calculated by vapor and liquid loading conditions at that point.

Vapor load consists of the water vapor (steam) and stripping gas flowing up through the still.

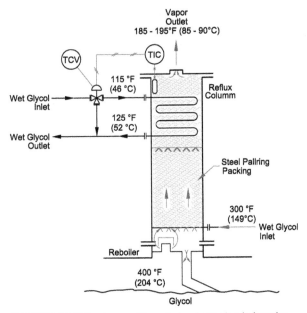

FIGURE 2-50 Still column with wet glycol entering below the stainless steel pall rings.

Liquid load consists of the rich glycol stream and reflux flowing downward through the still column.

The diameter required for the still is based on the glycol circulation rate (Figure 2-51).

Packing

1 to 3 theoretical trays (4 to 12 feet) is sufficient for most TEG stripping still requirements.

304 SS packing is normally used.

Amount of Stripping Gas

The amount of stripping gas required to reconcentrate the glycol to a high purity will range from 2 to 10 ft.^{3}/gal TEG circulated (Figure 2-52).

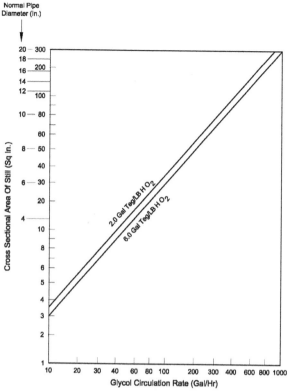

FIGURE 2-51 Determination of stripping still column diameter.

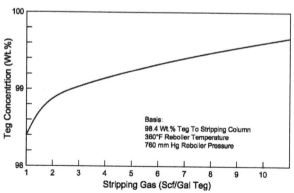

FIGURE 2-52 Amount of stripping gas required to reconcentrate glycol to high purity.

Filters

Microfiber
Sized to remove 5 micron solids.

Activated Charcoal (Carbon)
Used to remove chemical impurities.

Sized for full flow with 10 gpm streams.

Sized for 10 to 25% side streams on large units.

Glycol Pumps

Two types of pumps are used:

Glycol-gas powered pump

Electric driven positive displacement piston/plunger pump

Glycol-Gas Powered Pump (Figure 2-53)
Powered by gas entrained in the wet glycol leaving the contactor.

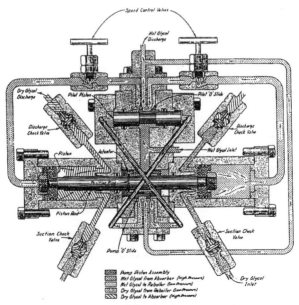

FIGURE 2-53 Operation of the glycol-gas powered pump (Kimray).

Does not require contactor glycol liquid level control, dump valve, or external power (electricity).

Gas consumption is relatively low and when used in conjunction with a glycol hydrocarbon skimmer or flash tank, very little gas loss is experienced.

Have few moving parts, which translates into less wear and simplified repairs.

Contact with hydrocarbon distillate, which may be entrained in glycol passing through the pump, swells o-ring seals in the pumps causing premature pump failure.

Generally used on small isolated systems

Inexpensive, break often, and are easy to repair

Circulation rate and gas consumption of the Kimray glycol-gas powered pump are shown in Tables 2-5 and 2-6.

Temperatures above 200°F damage o-ring seals.

Electric Driven Positive Displacement Piston/Plunger Pump

Usually used in large installations

Require a small glycol leak in the piston rod packing for lubrication.

Resilient to hydrocarbon distillate, grit, and debris that would damage the glycol-gas powered pumps.

Still Emissions

Vapor from the still column can contain some hydrocarbon gases that flashed from the glycol, stripping gas and aromatics.

Glycol preferentially absorbs aromatics and napthene components over paraffinic components in the inlet gas.

Aromatics:

Include benzine, ethylene, toluene, and xylene (commonly called BETX)

Condense with water vapor

Could lead to "soluble" oil in the produced water discharge

Treatment consists of condensing the water vapor and BETX exiting the still column and then compressing the noncondensables (hydrocarbon gases) (Figure 2-54)

Table 2-5 Glycol-Gas Circulation Rate (gallons/hour)

Model Number	Pump Speed (strokes/min) (Count One Stroke for Each Discharge of Pump)																
	8	10	12	14	16	18	20	22	24	26	28	30	32	34	36	38	40
1715V		10	12	14	16	18	20	22	24	26	28	30	32	34	36	38	40
4015V			12	14	16	18	20	22	24	26	28	30	32	34	36	38	40
9015V	8		27	31.5	36	40.5	45	49.5	54	58.5	63	67.5	72	76.5	81	85.5	90
21015V		66	79	92	105	118	131	144	157	171	184	197	210				
45015V		166	200	233	266	300	333	366	400	433							

NOTE: Do not attempt to run pumps at speeds less or greater than those indicated above.

Table 2-6 Gas Consumption

Operating Pressure (psig)	300	400	500	600	700	800	900	1000	1100	1200	1300	1400	1500
Gas Consumption (SCF/Gallon)	1.7	2.3	2.8	3.4	3.9	4.5	5.0	5.6	6.1	6.7	7.2	7.9	8.3

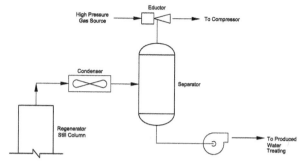

FIGURE 2-54 Process flow diagram of treatment of still emissions.

MERCURY CONSIDERATIONS

Mercury

Can be present in the formation as elementary mercury or can be introduced into the gas stream from instruments.

Has an affinity for higher molecular weight components.

Most stay with liquids rather than with the gas stream.

Reacts with iron oxide in the presence of hydrogen (H_2) and deposits mercurous sulfide on carbon steel pipe walls.

In the presence of condensed water combines with aluminum to form a weak amalgam.

Has a cumulative effect, thus even trace amounts can be harmful.

$$AL + Hg \rightarrow ALHg$$

and

$$2ALHg + 6H2O \rightarrow 2AL\,(OH)3 + 3H2 + 2Hg$$

Treatment

Treated with a sulfur-impregnated activated carbon with at least a 10% sulfur content.

System design considerations:

> Bed absorption = 15 to 20 weight % H_g
>
> Pressure range = 300 to 1100 psi
>
> Temperatures = up to 175°F
>
> Gas contact time = 20 seconds
>
> Maximum velocity = 35 fpm
>
> Regeneration = No commercial process

Proprietary treatment with alumina beds that can be recycled is offered by Rhone-Puolene.

SPECIAL GLYCOL DEHYDRATION SYSTEMS

General Considerations

When large dew-point depressions are required, specialized dehydration systems using highly concentrated glycol will be necessary.

If limited space is available, special systems may be used to achieve the required dew point depressions.

Normal dehydration systems with TEG glycol purity of 98.5% are capable of achieving dew point depressions up to 70°F.

Stripping gas may be used to obtain higher dew point depressions.

Vacuum-operated glycol units can achieve glycol purities of up to 99.9% but are rarely used because of:

> High operating costs
>
> Problems associated with achieving the necessary vacuums

The Drizo (wt.-2) and Cold Finger condenser process are other methods used to obtain low dew points.

Drizo (wt.-2) Process

Used to obtain glycol concentrations (purity) as high as 99.99% and low dew points in the range of −40°F to −80°F.

Uses an 80 to 100 molecular weight solvent (usually 150 octane) in the reconcentrator to form an azeotrope with water, thus lowering the effective boiling point of the mixture.

Glycol of higher purity is thus achieved for a given reconcentrator temperature.

BETX are collected as "excess" solvent.

Economic considerations:

May be favored over stripping gas.

Existing units can be retrofitted to increase the dehydration capacity.

Each situation must be evaluated on a case-by-case basis since Drizo (wt.-2) is a Dow-patented process and a license fee is required.

Process Description (Figure 2-55)

As shown in Figure 2-55, the Drizo process is the same as a conventional TEG dehydration system until the wet glycol flows into the reconcentrator.

Wet glycol is reconcentrated to 98.5% by conventional distillation.

The semi-lean glycol is then counter currently contacted with hydrocarbon solvent (iso-octane) vapors at 400°F.

The hydrocarbon and water are taken overhead, condensed, and then phase separated.

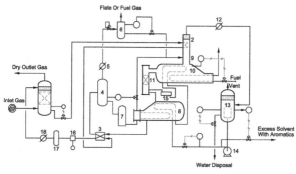

Process Flow Diagram Showing :

1. Glycol Contactor	7. Glycol Filter	13. Solvent-Water Separator
2. Reflux Condenser	8. Surge Tank/Exchanger	14. Solvent Pump
3. Glycol-Glycol Plate Exchanger	9. Rich Stripper	15. Solvent Superheater
4. Flash Tank	10. Glycol Reboiler	16. Glycol Pump
5. Solvent Recovery Condenser	11. Lean Stripper	17. Acoustical Filter
6. Recovered Solvent Drum	12. Solvent-Water Condenser	18. Glycol Cooler

FIGURE 2-55 Dow drizo (wt.-2) gas dehydration process.

The water is discarded, and the solvent is recycled into the system.

Applications

Competitive with applications that utilize a conventional TEG unit with stripping gas.

Most competitive in the range of −40°F to −80°F.

Should not be considered for hydrocarbon dew points.

Cold Finger Condenser Process

Based on the water TEG, vapor liquid equilibrium diagrams shown in Figure 2-56.

This diagram shows that for any liquid concentration the corresponding equilibrium vapor concentration is richer in water.

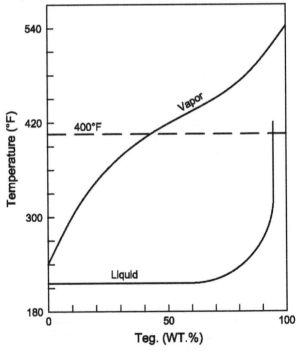

FIGURE 2-56 Vapor liquid equilibrium diagram for TEG.

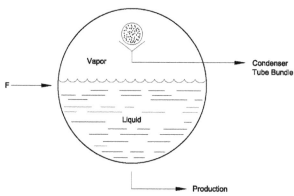

FIGURE 2-57 Cold finger condenser.

Incorporates a closed vessel one-half filled with vapor and liquid at equilibrium with a condenser tube bundle in the vapor space (Figure 2-57).

The condenser causes water condensation, which is removed from the vessel to a trough placed under the condenser tube bundle.

As the condensate is removed:

System's equilibrium is upset

Liquid phase releases more water to the vapor in order to reestablish equilibrium

Consequently, the liquid phase has a lower water content than it did originally.

Process Description

Numerous variations based on this principle exist.

One design is shown in Figure 2-58.

Contact between gas and glycol is the same as in a conventional TEG system.

Wet glycol leaves the contractor and flows to the condenser-tube bundle of the cold finger, where it:

Acts as a coolant

Is used as a coolant in the glycol still before the hydrocarbon liquid phase, hydrocarbon vapor phase, and glycol/water phase are separated in a three-phase separator

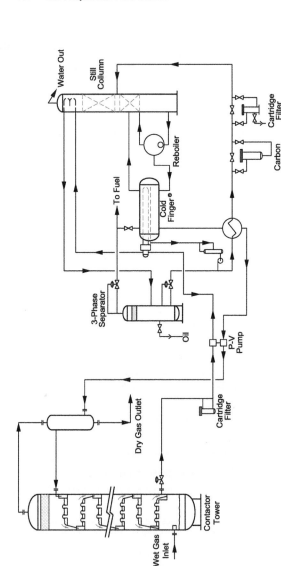

FIGURE 2-58 Cold finger condenser process.

The glycol/water phase is mixed with the cold finger condensate, and is heated by the cold finger liquid product before it is fed to the still.

The hot semi-lean glycol (which is near its boiling point) from the still bottoms is fed to the cold finger.

The liquid product is cooled, pumped, cooled again, and fed to the contactor.

Application

The main benefit of this system is that it is more fuel-efficient then the conventional TEG system.

However, it is more complex and not as well-proven as the conventional system.

SYSTEMS UTILIZING GLYCOL-GAS POWERED PUMPS (FIGURE 2-59)

Cool wet glycol leaves the bottom of the contactor, passes through a strainer, and powers the pump.

The wet glycol takes a pressure drop through the pump, then passes through the reflux condenser coil in the reconcentrator still column.

SYSTEMS UTILIZING ELECTRIC DRIVEN PUMPS (FIGURE 2-60)

Cool wet glycol leaves the bottom of the contactor, passes through a choke and level control operated motor valve, where a pressure drop occurs.

The glycol then passes through the still column reflux coils.

From the reflux coil it flows through the first dry glycol to wet glycol heat exchanger, then into the gas/glycol/condensate separator, where insoluble hydrocarbons are removed.

From the separator, the glycol passes through the filter to remove tarry hydrocarbons, then through the second dry glycol to wet glycol heat exchanger and into the reconcentrator still column.

In the top of the reboiler still column, cool wet glycol flows through the reflux condenser coils, preventing glycol from leaving as a vapor.

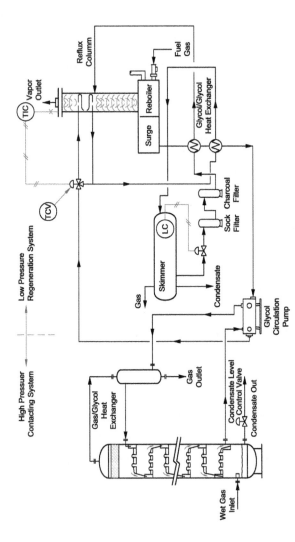

FIGURE 2-59 System utilizing glycol-gas powered pumps.

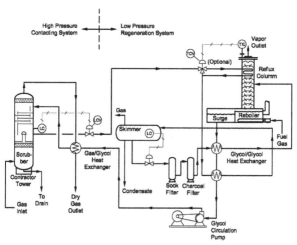

FIGURE 2-60 Systems utilizing electric driven pumps.

The wet glycol enters the column below the coils and spills downward through the packing and into the reconcentrator.

Heat is circulated through the tube to boil the water from the glycol.

A weir holds the level of glycol above the heating tubes.

Regenerated glycol flows over the weir and leaves through the outlet in the bottom.

Example 2-3: Glycol Dehydration

Given:

Gas $Q_g = 98$ MMSCFD at 0.67 SG saturated with water at 1000 psig and 100°F

Dehydrate to $= 7$ lb/MMSCF

Use triethylene glycol

No stripping gas is available

98.5% TEG concentration

C_D (contactor) $= 0.852$

$T_c = 376°R$

$P_c = 669$ psia

Determine:

1. Calculate contactor diameter

2. Determine glycol circulation rate and estimate reboiler duty

3. Size the still column

4. Calculate heat duties for gas/glycol exchanger and glycol/glycol exchangers

Solution:

1. Calculate contactor diameter

$$d^2 = 5040 \left(\frac{TZQ_g}{P} \left[\left(\frac{\rho_g}{\rho_L - \rho_r} \right) \frac{C_D}{d_M} \right] \right)^{1/2}$$

$d_M = 125$ microns (range 120–150 microns)

$T = 570°R$

$P = 1015$ psia

$Q_g = 98$ MMSCFD

$T_r = 570/376 = 1.49$

$P_r = 1015/669 = 1.52$

$Z = 0.865$

$$\rho_g = \left[\frac{(0.67)(1015)}{(560)(0.865)} \right]$$

$$= 3.79 \text{ lb/ft.}^3$$

$\rho_L = 70 \text{ lb/ft.}^3$

$C_D = 0.852$ (given)

$$d^2 = 5040 \frac{(560)(0.685)(98)}{(1015)} \left[\left(\frac{3.72}{70 - 3.79} \right) \frac{0.852}{125} \right]^{1/2}$$

$$= 68.2 \text{ in}$$

Use 72″ ID contactor (standard off-the-shelf)

2. Determine glycol circulation rate and reboiler duty

$W_i = 63$ lb/MMSCF (from McKetta-Wehe) (saturated water content)

$W_O = 7$ lb/MMSCF (spec)

$\Delta W = W_i - W_O = 63 - 7 = 56$ lb/MMSCF

$\Delta W/W_i = 56/63 = 0.889$

Using $n = 2$ (i.e., 8 actual trays) and glycol purity of
98.5% read from Figure 2-48 the glycol circulation rate
of 2.8 gal TEG/lb H_2O. Use 3.0 gal/lb for design.

$$L = \left(\frac{2.0 \text{ gal}}{\text{lb}}\right)\left(\frac{56 \text{ lb}}{\text{MMSCF}}\right)\left(\frac{98 \text{ MMSCF}}{D}\right)\left(\frac{D}{24 \text{ hr}}\right)\left(\frac{\text{hr}}{60 \text{ min}}\right)$$

$= 11.4$ gpm TEG

$= 862$ Btu/gal(table)

$$= \left(\frac{862 \text{ Btu}}{\text{gal}}\right)\left(\frac{11.4 \text{ gal}}{\text{min}}\right)\left(\frac{60 \text{ min}}{\text{hr}}\right)$$

$= 590$Mbtu/hr

To allow for start-up heat loads, increase heat duty by
10% and then select a standard off-the-shelf fire tube.
Thus, select a 750 MMBtu/hr.

3. Design of still column:

Use 12-foot still column (standard packed
arrangement)

$d_M = 125$ micron

$T = 300°F = 760°R$

$P = 1$ psig

$$Q_g = \left(\frac{10 \text{ scf}}{\text{gal}}\right)\left(\frac{11 \text{ gal}}{\text{min}}\right)\left(\frac{60 \text{ min}}{\text{hr}}\right)\left(\frac{24 \text{ hr}}{d}\right)$$

$= 0.16$ MMSCFD

$Z = 1.0$

$$\rho_g = 2.7\left(\frac{(0.62)(16)}{(760)(1.0)}\right)$$

$= 0.035$ lb/ft.^{3}

$\rho_L = 62.4$ lb/ft.^{3}

$C_D = 14.2$ (given)

$$d2 = 5040\frac{(760)(1.0)(0.16)}{16}\left[\left(\frac{0.035}{62.4 - 0.035}\right)\frac{14.2}{125}\right]^{1/2}$$

$= 17.5$ inches

Use 18 inch OD x 12 feet long still

4.   Calculate duties of heat exchangers

Rich TEG from contactor: $T = 100°F$ (given)

Rich TEG to separator: $T = 200°F$ (assume for good design)

Rich TEG from reflux: $T = 110°F$ (assume 10°F increase in reflux coil)

Rich TEG to still: $T = 300°F$ (assume for good design)

Lean TEG from reboiler: $T = 385°F$ (from Figure 2.59)

Lean TEG to pumps (max): $T = 210°F$ (from manufacturer)

Lean TEG to contactor: $T = 110°F$ (10°F above contactor temperature)

5.   Glycol/glycol preheater (rich side, duty):

Rich TEG: $T_1 = 110°F$ (assume 10°F increase in reflux coil)

$T_2 = 200°F$

Lean glycol composition:

$$W_{TEG} = (0.985)\left(\frac{70\ lb}{ft.^3}\right)\left(\frac{ft.^3}{7.48\ gal}\right)$$

$$= 9.22\ lb\ TEG/gal\ of\ lean\ glycol$$

$$W_{H_2O} = (0.015)\left(\frac{70\ lb}{ft^3}\right)\left(\frac{ft^3}{7.48\ gal}\right)$$

$$= 0.140\ lb\ H_2O/gal\ of\ lean\ glycol$$

Rich glycol composition:

$$W_{TEG} = 9.22\ lb\ TEG/gal\ of\ lean\ glycol$$

$$W_{H_2O} = \left(\frac{0.140\ lbH_2O}{gal\ of\ lean\ glycol}\right) + \left(\frac{1\ lbH_2O}{3.0\ gal\ of\ lean\ glycol}\right)$$

$$= 0.473\ lb\ H_2O/gal\ of\ lean\ glycol$$

$$\text{Wt.Concentration TEG} = \frac{9.22}{9.22 + 0.473}$$

$$= 95.1\%$$

Rich glycol flow rate (wrich):

$$W_{rich} = (9.22 + 0.473)\frac{lb}{gal}\left(\frac{11.4\ gal}{min}\right)\left(\frac{60\ min}{hr}\right)$$

$$= 6630\ lb/hr$$

Rich glycol heat duty (qrich):

C_P (95.1% TEG) = 0.56 at 110°F (from physical property of TEG) and 0.63 at 200°F = 0.60 Btu/hr°F

$C_{P,AVG}$ = 0.60 Btu/hr°F

$$q_{rich} = \left(\frac{6630\ lb}{hr}\right)\left(\frac{0.6\ btu}{hr}\right)(200 - 110)°F$$

$$= 358\ MBtu/hr$$

6. Glycol/glycol exchanger

Rich T_1 = 200

T_2 = 300

Lean T_3 = 390

T_4 = ?

Rich glycol heat duty:

C_P (95.1% TEG) = 0.63 at 200°F (from physical properties of TEG) = 0.70 at 300°F

$C_{P,AGV}$ = 0.67 Btu/hr°F

$$q_{rich} = \left(\frac{6630\ lb}{hr}\right)\left(\frac{0.67\ Btu}{lb°F}\right)(300 - 200)°F$$

$$= 444\ MBtu/hr$$

Lean glycol flow rate (Wlean):

$$W_{lean} = \left(\frac{11.4\ gal}{min}\right)\left(\frac{70\ lb}{ft^3}\right)\left(\frac{ft^3}{7.48\ gal}\right)\left(\frac{60\ min}{hr}\right)$$

$$= 6401\ lb/hr$$

Calculation of T4

Assume T = 250°F

T_{AVG} = (353 + 250)/2 = 302°F

$C_{P,AGV}$ = (98.5% TEG) = 0.67 Btu/lb°F (from physical properties of TEG)

$$Q_{lean} = W_{lean}\,C_p\,(T_4 - T_3)$$

$$Q_{lean} = -q_{rich}$$

$$T_4 = T_3 - \left(\frac{q_{rich}}{W_{lean}C_p}\right)$$

$$= 353 - \left(\frac{444,000}{(6401)(0.67)}\right)$$

$$= 249°F$$

Temperature:

Lean: $T_4 = 249°F$

$T_5 = ?$

Assume $T_5 = 175°F$

$T_{AV} = (249 + 175)/2 = 212$

$C_{P,AV} = (98.5\% \text{ TEG}) = 0.61$ Btu/lb°F (from physical properties of TEG)

$$q_{lean} = W_{lean}\,C_P\,(T_4 - T_5)$$

$$q_{lean} = -q_{lean}$$

$$T_5 = T_4 - \left(\frac{q_{rich}}{W_{lean}C_p}\right)$$

$$= 249 - \left(\frac{358,000}{(6401)(0.61)}\right)$$

$= 157°F$ (this is less than the maximum allowed to the pumps)

Lean: $T_1 = 157°F$

$T_2 = 110°F$

C_P (98.5% TEG) $= 0.57$ at 157°F (from physical properties of TEG) $= 0.53$ at 110°F

$C_{P,AGV} = 0.55$ Btu/lb°F

$q_{lean} = (6401)\,(0.55)\,(110{-}1\,57) = -165$ MBtu/hr

Glycol/glycol exchanger:

Rich $T_1 = 200°F$, $T_0 = 300°F$

Lean $T_2 = 353°F$, $T_0 = 249°F$

Duty $q = 444$ MBtu/hr

NONREGENERABLE DEHYDRATOR

Overview

One other general category will be covered specifically, the calcium chloride brine dehydrator.

Calcium Chloride Unit

Calcium chloride ($CaCl_2$) dehydrator is the most common (Figure 2-61).

Unit consists of three sections:

Inlet gas scrubber

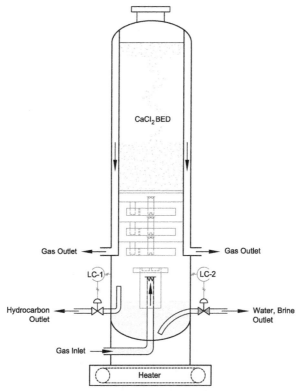

FIGURE 2-61 Cross section of Calcium Chloride dehydrator.

Brine tray

Solid brine particles

Only moving parts are the liquid level
controls for hydrocarbon liquid and the brine-water
mixture.

Principles of Operation

Solid desiccant is placed in the top of the unit.

Water-wet gas containing the solid $CaCl_2$ gives up
part of its water to form liquid brine to drip down
and fill the trays.

Inlet gas coming up through the specially
designed nozzles on the trays contact the brine
efficiently.

The wettest gas contacts the most dilute brine
(about 1.2 specific gravity).

Approximate 2.5 Lb H_2O/lb $CaCl_2$ is removed in the
trays.

Brine gravity on the top tray is about 1.4.

Another 1 lb H_2O/lb $CaCl_2$ is removed in the solid
bed section.

Maximum dew point depression of 60° to 70°F
occurs in this section.

Typically used in remote, small gas fields without
heat or fuel.

Advantages

Simple

No moving parts

No heat required

Does not react with H_2S or CO_2

Can dehydrate hydrocarbon liquids

Disadvantages

Batch process

Emulsifies with oil

Unreliable

Limited dew point depression

Brine disposal is a problem

Sensitive to varying flow rates

Operating Problems

Bridging and channeling is a problem.

Brine can crystallize at 85°F, thus during low flow periods can plug vessel outlet or trays.

Brine carry-over can cause severe corrosion problems.

Design Considerations

Figure 2-62 illustrates the water content of natural gas dried by solid calcium chloride bed units.

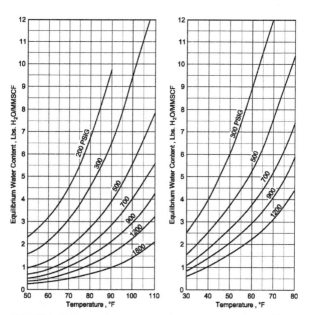

FIGURE 2-62 Approximate water content on natural gas driven by $CaCl_2$ unit (*Left*: freshly recharged; *right*: just prior to recharging).

PHYSICAL PROPERTIES OF COMMON GLYCOLS

Figures 2-63 through 2-73 contain specific heats, specific gravities and viscosities of EG, DEG, TEG and TTEG solutions.

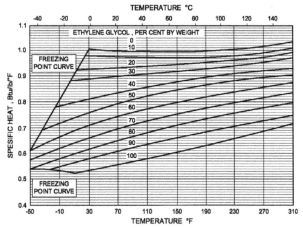

FIGURE 2-63 Specific heat of aqueous EG solutions.

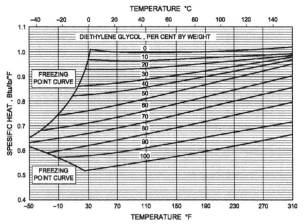

FIGURE 2-64 Specific heat of aqueous DEG solutions.

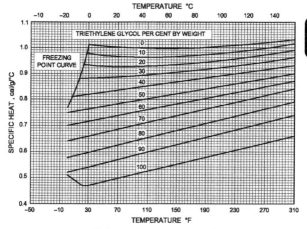

FIGURE 2-65 Specific heat of aqueous TEG solutions.

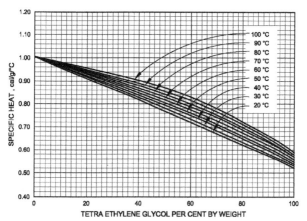

FIGURE 2-66 Specific heat of aqueous TTEG solutions.

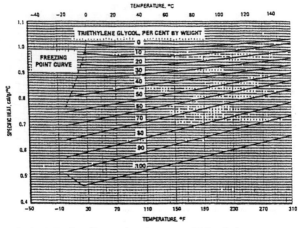

FIGURE 2-67 Specific gravity of aqueous TEG solutions.

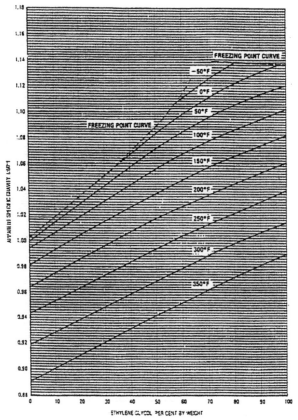

FIGURE 2-68 Specific gravity of aqueous EG solutions.

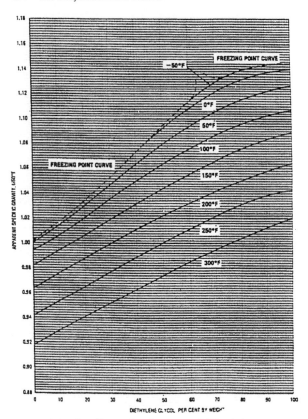

FIGURE 2-69 Specific gravity of aqueous DEG solutions.

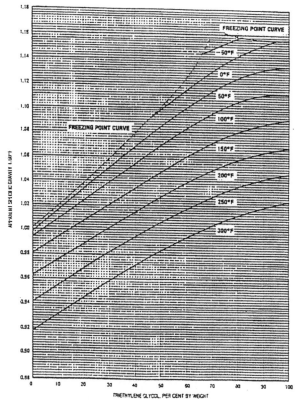

FIGURE 2-70 Specific gravity of aqueous TEG solutions.

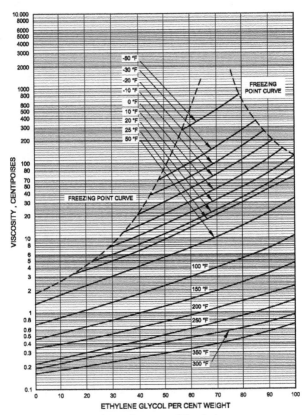

FIGURE 2-71 Viscosities of aqueous EG solutions.

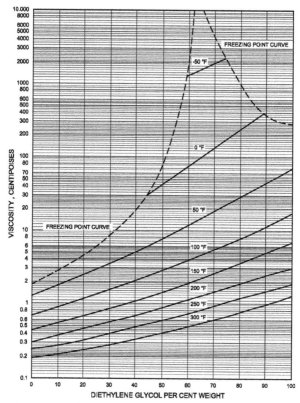

FIGURE 2-72 Viscosities of aqueous DEG solutions.

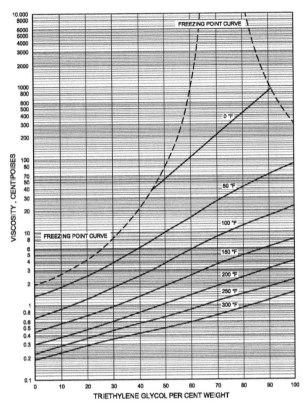

FIGURE 2-73 Viscosities of aqueous TEG solutions.

Part 3
Glycol Maintenance, Care, and Troubleshooting

3

Contents

PREVENTIVE MAINTENANCE

Scheduled Preventive Maintenance

Scheduled preventative maintenance reduces glycol losses such as:

Foaming

System plugging

It reduces mechanical failure such as:

Corrosion

Pump failures

It also minimizes system down time.

It maximizes system operation efficiency.

Five Steps to a Successful Preventive Maintenance Program

Record-Keeping

Accurate records can be used to determine the system efficiency and to pinpoint operating problems.

Records of prior and existing conditions including dew points, glycol usage, and repairs help establish the *system profile*.

DOI: 10.1016/B978-1-85617-980-5.00003-3

Once the system profile is defined it becomes easier to identify unusual system characteristics that may indicate potential problems.

Mechanical Maintenance

Daily physical inspections are necessary to insure that the system is running properly.

Any trouble encountered should be dealt with immediately, thus preventing the problem from escalating.

Glycol Care

Regular chemical analysis (every one or two months) of the glycol provides detailed information on the internal operation of the unit.

Many process-related problems can be diagnosed well in advance of mechanical failure.

Chemical problems can be diagnosed and corrective action taken before they become costly and detrimental to unit performance.

Corrosion Control

Corrosion is a frequent problem in glycol dehydration systems (see Figures 3-1 through 3-4).

FIGURE 3-1 Oxidation corrosion of bubble cap tray.

FIGURE 3-2 Sweet corrosion of valve tray.

If unchecked the damage can be extensive.

All units should have provisions for corrosion control.

Communication

Lines of communication between field and office personnel are critical to the smooth operation of any system.

Office personnel (production supervisors, facility engineers, purchasing agents) must be kept informed of daily operations and any problems that may arise.

Field personnel must be made aware of technical information that may improve their operations.

FIGURE 3-3 Sour corrosion of bubble caps.

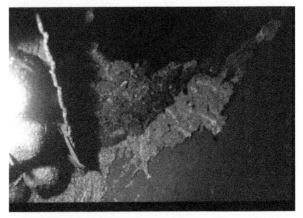

FIGURE 3-4 Generalized oxidation corrosion of contact tower.

Training for field operators allows the operator to better maintain the equipment.

Record-Keeping

Records necessary to establish a system profile include:

Design information including vessel specifications, equipment drawings, and P&IDs

Filter element or media replacement—type and frequency

Glycol usage—gallons/month

Chemical additives—type and amount

Gas production and flow rate charts—peak, average, and low periods

3

Outlet gas dew point/water content (lbs/MSMCF)

Mechanical inspections—type, magnitude, frequency, results

Records necessary to establish a system profile include:

Glycol analysis—format, frequency, recommendations, results

Corrosion coupon results—mills per year (MPY), frequency

Materials and labor relating to system repairs—operating costs

Figures 3-5, 3-6, and 3-7 are examples of common report formats.

With the aforementioned information, a good system profile can be drawn of a specific system.

Updating these records will show any gradual changes in a unit's system profile and may alert you to a potential or developing problem.

Mechanical Maintenance

The following things should be done so as to keep the unit operating properly and to prevent operational problems:

1. Insure that instruments and controls are in good working condition (thermometers and pressure gauges, etc.). Use a test thermometer on the reconcentrator to insure proper reconcentrator heat.

2. Insure glycol filter elements are changed according to average expected need basis:

    Microfiber filters should be changed monthly.

    Carbon filters should be changed monthly (small cartridge filters) to every six months (large bulk units).

    Glycol analysis helps determine the frequency.

Monthly Glycol Dehydration Report

COMPANY ____________________ MONTH & YEAR ____________________

LOCATION ____________________ GLYCOL PUMP TYPE/MODEL ____________________

DAY	REBOILER TEMP.	STILL TEMP.	PUMP SPEED STROKES/MIN	GLYCOL ADDED GALLONS	GAS RATE DAILY MMSCF	GLYCOL TEMP. PUMP SUCTION		CONTACT TOWER GAS PARAMETER		SOCK DIFF.	OPERATOR	COMMENTS
						LEAN	RICH	TEMP	PRESS.			
1												
2												
3												
4												
5												
6												
7												
8												
9												
10												
11												
12												
13												
14												
15												
16												
17												
18												
19												
20												
21												
22												
23												
24												
25												
26												
27												
28												
29												
30												
31												

FIGURE 3-5 Monthly glycol dehydration report.

Filter Change Schedule

COMPANY ____________________ SOCK TYPE & QUANTITY ____________________

LOCATION ____________________ CARBON TYPE & QUANTITY ____________________

Sock Filters		Carbon Filters		Operator	Comments
Date Changed	Inventory	Date Changed	Inventory		

FIGURE 3-6 Filter change schedule.

Monthly Chemical Usage

COMPANY ______________ MONTH & YEAR ______________

LOCATION ______________ GLYCOL CAPACITY ______________

	GLYCOL		CORROSION INHIBITOR		DE-FOAM INHIBITOR		COMMENTS
DAY	QUANTITY ADDED IN GALLONS	INVENTORY	QUANTITY IN GALLONS	INVENTORY	QUANTITY IN GALLONS	INVENTORY	
1							
2							
3							
4							
5							
6							
7							
8							
9							
10							
11							
12							
13							
14							
15							
16							
17							
18							
19							
20							
21							
22							
23							
24							
25							
26							
27							
28							
29							
30							
31							

FIGURE 3-7 Monthly chemical usage.

An upset or sudden change in the operating conditions may foul the filters faster than the preventive maintenance program anticipates.

Make sure filter differential pressure is below 15 psi.

3. Look for glycol leaks on and around the glycol skid.

   Most leaks can be stopped by tightening a union, valve stem packing, or pump rod packing.

   After the leak has been repaired, clean the affected area so it is easier to notice new leaks.

4. Check the glycol level, at least twice a day, and add glycol as necessary.

   Maintain a written report of glycol added.

   This allows operations to detect excessive losses of glycol and take corrective action faster.

5. Insure unit performance by taking a dew point measurement daily.

6. Clean the glycol strainers monthly to prevent accumulation of trash, which can cause the glycol pump to fail.

7. Check the glycol circulation rate daily.

   Any time the gas flow rate changes or when a drastic change in gas pressure or temperature is experienced, the glycol flow rate should be recalculated and the pumps set accordingly.

   On multiple pump installations, switch the pumps weekly, thus insuring pump operation when necessary.

8. On direct fired fire tubes, sight down the fire tube weekly for fire tube blisters or hot spots.

   These indicate extreme fouling and impending fire tube failure.

9. Cycle the main burner manually to be sure the fuel gas valve works and the pilot light stays lit.

   Check the fuel gas scrubber pot for fluid build-up that may hinder burner operation.

Glycol Care

General Considerations

Operating and corrosion problems occur when the circulating glycol gets dirty.

Some contaminated glycol problems can be noticed easily and corrective action taken.

A small glycol sample should be taken daily from the surge tank or dry glycol suction header to the pump.

Check closely for fine black particles settling out of the sample, which may be corrosion by-products and indicate an internal corrosion problem (see Figures 3-8 and 3-9).

Smell the Sample

If the smell is sweet and aromatic (similar to rotten bananas) it may be thermally decomposed.

If the sample is viscous and black it is probably contaminated with hydrocarbon or well-treating chemicals.

If the hydrocarbon contamination is great enough, the sample will separate into two liquids or *interphase*.

Every one or two months send a sample of both the rich and lean glycol to a laboratory for complete analysis.

FIGURE 3-8 Sample of uncontaminated glycol.

FIGURE 3-9 Three samples of liquid-hydrocarbon contaminated glycol: (a) moderate; (b) severe; (c) slightly.

This type of analysis will provide a detailed description of unit performance and glycol condition.

Corrosion Control

Overview

General Considerations

Corrosion is a major cause of premature equipment failure.

Corrosion can occur over the entire system, inside and out.

The two most common areas of severe corrosion are:

Still column reflux coil

Vent/fill connection on the surge tank

This is due to a high concentration of water vapor in the top of the still and the ready availability of oxygen in the air at the vent/fill cap.

Three types of corrosion that are almost always in glycol systems either individually or in combination with one another are:

Oxidation (Figure 3-1)

Sour corrosion (Figure 3-3)

Sweet corrosion (Figure 3-2)

Oxidation

Oxidation of metal is the exchange of electrons between metal and oxygen molecules to form positive and negative hydrogen ions.

Some metal loss is incurred.

The resultant scale-like residue formed by the process is called oxide, or *rust*.

Oxidization is characterized by rough, irregular, shallow pitting of the metal scaled over the rust.

Sour Corrosion

Acid gases (H_2S and CO_2) are often found in produced natural gas.

Glycols are very reactive with sulfur compounds, such as H_2S, and will exchange electrons with metal molecules, which initiate corrosion.

The resulting materials tend to polymerize (form larger molecules) that form a "gunk" that is very corrosive.

Sour corrosion is characterized by deep, jaggered pitting.

Sweet Corrosion

Water is found in glycol as vapor, free condensed water, or entrained in glycol.

Carbon dioxide (CO_2) when dissolved in water forms carbonic acid.

Since most produced natural gases contain some CO_2, the presence of carbonic acid in glycol systems is very common.

The corrosion resulting from carbonic acid is known as *sweet corrosion*.

Sweet corrosion is characterized by deep, round, smooth pitting.

Sometimes the pitting will cover a broad area, disguising the depth of the pit.

Prevention and Control Programs
General Considerations

Prevention and control programs should include system monitoring through:

Corrosion coupons

Glycol analysis (pH and iron)

3

Three steps in combating corrosion in glycol systems are:

Use an effective corrosion inhibitor in both the liquid and vapor phases.

Use corrosion resistant alloys (CRA) in construction.

Keep the unit clean to prevent acid formation due to contamination.

Cathodic protection has been attempted but met with little success.

It is impractical to attempt to eliminate corrosion completely.

The rate of corrosion can be slowed to a point that is almost negligible.

The maximum acceptable corrosion rate is 6 mils per year (MPY).

Corrosion inhibitors work in several ways.

The two most applicable to glycol units are:

pH buffers

Plating inhibitors

pH Buffers

pH buffers include:

Alkanolamines

MEA

TEA

The fight corrosion by stabilizing the pH near neutral, thereby reducing corrosive environment.

Amines are not true inhibitors in that no protection is actually afforded the metal surface.

Alkanolamines are regenerable as is the glycol and thus can be retained in the system for lengthy time periods.

However, they are thermally degraded at normal reconcentrator operating temperatures and if used frequently may leave harmful residues within the system.

Plating Inhibitors

Tallow diamine, unlike the inorganic amines like alkanolimines, is an organic amine.

It is grouped with the plating inhibitors even though it does not actually plate out on the vessel walls.

It flashes out of the glycol at high temperatures.

As it vaporizes it contacts the vapor spaces of the reconcentrator and forms a tenacious film over the exposed metal.

This film will eventually wear away and must be replenished occasionally to continue protection.

To a lesser degree than the alkanolamines, it will also buffer pH.

True plating inhibitors include:

Borax

NaCap (Sodium mercaptobenzothiazole)

Dipotassium phosphate

These inhibitors are strictly liquid phase protection.

They will plate out on the vessel walls forming a protective barrier between a corrosive environment and the metal.

This barrier also prevents the ready exchange of electrons, which drastically slows corrosion.

Since the plating inhibitors are all alkalies, some degree of pH buffering will be effected.

The pH buffering will not be as great as through the use of amines.

Communication

Communication is the easiest portion of an effective maintenance program and yet it is the most overlooked.

Communication can be between:

> Management and labor
>
> Engineer and foreman
>
> Operators on opposite shifts
>
> Office and field personnel

Lack of communication is the single most contributing factor to glycol system failure.

> When were the glycol filters changed?
>
> How much glycol losses are being experienced?
>
> What are the results of glycol analysis?
>
> What is the immediate history of the problem?

Failure to communicate can cause confusion and evolve into major problems.

General Considerations

Operating and corrosion problems usually occur when the circulating glycol gets dirty.

To achieve a long, trouble-free life from the glycol, it is necessary to recognize these problems and know how to prevent them.

Some of the major areas are:

> Oxidation
>
> Thermal decomposition
>
> pH control
>
> Salt contamination
>
> Hydrocarbons
>
> Sludge
>
> Foaming

Oxidation

Oxygen enters the system with the incoming gas through:

> Unblanketed storage tanks and sumps
>
> Pump packing glands

Sometimes glycol will oxidize in the presence of oxygen and form corrosive acids.

To prevent oxidation:

> Bulk storage tanks should be gas blanketed.
>
> Use oxidation inhibitors.
>
> Normally, a 50/50 blend of MEA and 33⅓% hydrazine is inserted into the glycol between the absorber and the reconcentrator.

A metering pump should preferably be used to give continuous, uniform injection.

Thermal Decomposition

Excessive heat, a result of the following conditions, will decompose glycol and form corrosive products:

> High reconcentrator temperature above the glycol decomposition level
>
> High heat-flux rate, sometimes used by a design engineer to keep the heater cost low
>
> Localized overheating, caused by deposits of salts or tarry products on the reconcentrator fire tubes or by poor flame direction on the fire tubes

pH Control

pH is a measure of the acidity or alkalinity of a fluid, based on a scale of 0 to 14.

> pH values from 0–7 indicate the fluid is acidic.
>
> pH values from 7–14 indicate the fluid is alkaline.

To obtain a true reading, glycol samples should be diluted 50-50 with distilled water before pH tests are run.

The pH meter should be calibrated occasionally to keep it accurate.

The pH of the stilled water should also be checked to assure that it has the neutral value of 7.

New glycol has a neutral pH of approximately 7.

> With usage the pH decreases and the glycol becomes acidic and corrosive unless pH neutralizers or buffers are used.
>
> Equipment corrosion rate increases rapidly with a decrease in the glycol pH.

Acid created by glycol oxidization, thermal decomposition products, or acid gases picked up from the gas stream are the most troublesome of corrosive contaminants.

A low pH accelerates the decomposition of glycol.

Ideally, the glycol pH should be held at a level between 7.0 and 7.5.

3

> A value above 8.5 tends to make glycol foam and emulsify.

> A value below 6.0 corresponds to system contamination, corrosion, and/or oxidation.

Borax, ethanolamines (usually triethanolamine) or other alkaline neutralizers are used to control the pH.

> These neutralizers should be added slowly and continuously for the best results.

> An overdose will usually precipitate a suspension of black sludge in the glycol.

> The sludge could settle and plug the glycol flow in any part of the circulating system.

> Frequent filter element changes should be made while pH neutralizers are added.

> The amount of neutralizer to be added and the frequency will vary from location to location.

Normally, ¼ lb of triethanolamine (TEA) per 100 gallons of glycol is sufficient to raise the pH level to a safe range.

When the glycol pH is extremely low, the required amount of neutralizer can be determined by titration.

> For best results, the lean rather than the rich glycol should be treated.

It takes time for the neutralizer to mix thoroughly with all the glycol in the system.

Several days are required before the pH is raised to a safe level.

Each time that neutralizer is added, the pH of the glycol should be measured several times.

Salt Contamination

Salt Deposits

> Salt deposits accelerate equipment corrosion.

> It also reduces heat transfer in the fire tubes.

It alters specific gravity readings when a hydrometer is used to measure glycol water concentration.

It cannot be removed with normal regeneration.

A scrubber installed upstream of the glycol plant should be used to prevent salt carry-over from produced free water.

In areas where large quantities of brine are produced, some salt contamination will occur.

The removal of salt from the glycol solution is then necessary. The following reclaiming methods are used:

> Scraped surface heat exchangers in conjunction with centrifuges
>
> Vacuum distillation
>
> Ion exchange
>
> Ion retardation

Hydrocarbons

Liquid hydrocarbons, a result of carry-over with the incoming gas or condensation in the contactor, increases glycol by:

> Foaming
>
> Degradation
>
> Losses

It must be removed with:

> Glycol/gas/condensate separator
>
> Hydrocarbon liquid skimmer
>
> Activated carbon beds

Sludge

Solid particles and tarry hydrocarbons (sludge) are suspended in the circulating glycol, and with time will settle out (see Figure 3-10).

This action results in the formation of black, sticky, abrasive gum that can cause trouble in pumps, valves and other equipment, usually when the glycol pH is low.

FIGURE 3-10 Sludge captured on glycol microfiber filters.

Foaming

General Considerations

Excessive turbulence and high liquid-to-vapor contacting velocities usually cause the glycol to foam (this condition can be caused by mechanical or chemical problems).

The best way to prevent foaming is proper care of the glycol, such as:

> Effective gas cleaning ahead of the glycol system

> Good filtration of the circulating solution

Defoamers

Defoamers serve only as a temporary control until the conditions generating foam can be identified and removed.

Success depends on when and how it is added.

Some act as good inhibitors if added after the foam has been generated, but aggravate the problem if added prior to onset by serving to stabilize the foam.

Most are inactivated within a few hours under high temperature and pressure, and thus their effectiveness is dissipated by the heat of the glycol solution.

Thus, defoamers should be added continuously, a drop at a time, for best results.

Chemical feed pumps:

Meter the defoamer accurately

Improve dispersion into the glycol

Are activated automatically by differential pressure across the contactor

Analysis and Control of Glycol

General Considerations

Analysis of glycol is essential to good plant operation.

It helps pinpoint high glycol losses, foaming, corrosion, and other operating problems.

Analyses enable operations personnel to evaluate plant performance and make operating changes to obtain maximum drying efficiency.

Visual Inspection

A glycol sample should first be visually inspected to identity some of the contaminants (see Figure 3-11).

A finely divided black precipitate may indicate the presence of iron corrosion products.

A black, viscous solution may contain heavy, tarry hydrocarbons.

The characteristic odor of decomposed glycol (a sweet aromatic odor) usually indicates thermal degradation.

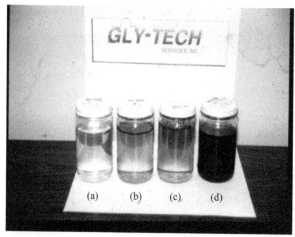

FIGURE 3-11 Glycol samples: (a) normal glycol; (b) hydrocarbon carry-over; (c) iron corrosion particles settling to the bottom of sample container; (d) two-phase large liquid hydrocarbon carry-over. (Courtsey of Gly-Tech)

A two-phase liquid sample usually indicates the glycol is heavily contaminated with hydrocarbons.

The visual conclusion should next be supported by a chemical analysis.

Chemical Analysis

A complete glycol analysis of lean and rich samples, when properly interpreted, can provide a detailed picture of the workings of the dehydration unit and its process.

Glycol analysis should include tests to determine the following (refer to Table 3-1):

> pH (50/50)
>
> Hydrocarbon content (% wt.)
>
> Water content (% wt.)
>
> Total suspended solids (% wt.)
>
> Residue (% wt.)
>
> Chlorides (mg/l)
>
> Iron (mg/l)

Foam character

Height (ml)

Stability (sec.)

Specific gravity

Glycol composition

EG

DEG

TEG

TTEG

Chemical Analysis Interpretation

pH

A pH below 6 generally corresponds with system contamination, corrosion, and/or oxidation.

Below 5.5 autoxidation occurs.

Chemical decomposition of the glycol occurs within itself.

Glycol has the tendency to continue to drop without outside influences.

Causes of low pH:

Acid gases in the gas stream

Organic acids due to oxidation or thermal degradation

Excessive chlorides (salt) in the glycol

Well-treating chemicals entrained in the gas stream

Thermal decomposition of entrained liquid hydrocarbons in the gas stream and glycol

Oxidation of the glycol due to improper storage

Causes of high pH:

Contamination from well-treating chemicals entrained in the gas stream

Overdose of neutralizer added to a system for low pH

Foaming tendencies can result from high pH, due to stabilized glycol-hydrocarbon emulsions.

Table 3-1 Chemical Analysis

Company _____   Date: _____
Location _____

Test	Lean Glycol	Rich Glycol	Allowable Range	Ideal
pH (50/50)			6 to 8	7 to 7.5
Hydrocarbon (% wt.)			0.1%	
Water content (% wt.)			2% lean 6% rich	
TSS (% wt.)			0.01%	
Residue (% wt.)			4%	2%
Chlorides (mg/l)			1500	1000
Iron (mg/l)			50	35
Foam character:				
Height (ml)			20 to 30 ml	
Stability (sec)			15 to 5 sec	
Specific gravity			1.118 to 1.126	
Glycol composition:				
EG				
DEG				
TEG				
TTEG				

Sludge and residue build-up can result from both high and low pH.

Sludge

May become abrasive and cause premature pump and valve failure.

May deposit in trays and downcomers, still column packing, and heat exchangers, which cause system plugging.

Hydrocarbons

Enter the glycol stream as a result of inlet separator carry-over or as condensation due to temperature variations.

Compressor lube oils and other extraneous organic chemicals such as pipeline corrosion inhibitors, are often stripped out of natural gas as it passes through the contact tower.

Oils and organic residues can cause glycol/water emulsions and suspensions, which contribute to foaming:

Results in excessive high glycol carry-over from the contactor

Contaminants may cause plugging in the contactor, still column, and heat exchangers

Light hydrocarbons:

Usually separated from the glycol stream with an adequately sized glycol/hydrocarbon separator

Heavy hydrocarbons:

Referred to as soluble hydrocarbons because they bond with the glycol

Usually filtered out with activated carbon

Light end hydrocarbons (insoluble) are allowable up to 1% by volume.

Soluble hydrocarbons are only acceptable to 0.1 % by weight.

Primarily responsible for foaming, sludge and residue build up, low pH, loss of hygroscopicity, and glycol decomposition.

Water Content

Water content is defined as the quantity of water in the glycol.

The difference between the lean sample and rich sample measures the degree of loading in the contactor.

It indicates regeneration efficiency.

Glycol purity should be at least 98% in the lean stream and at least 94% in the rich.

These concentrations will produce the desired dew points in systems that are operating properly.

For lower dew points the glycol purity must be increased (or water content decreased).

High water content of the lean sample generally indicates low reconcentrator heat.

High water content in the lean sample may also indicate:

Excessive glycol circulation

Undersized equipment

3

Carry-over from the separator

Vapor communication from reconcentrator to surge

A leak in the glycol/glycol heat exchanger

Over-refluxing in the still column

Hot inlet gas temperature

High water content in the rich sample usually indicates a low glycol circulation rate or:

Carry-over from the separator

Poor reconcentration

Heat exchanger communication

Undersized equipment

Hot inlet gas temperature

Check values for hydrocarbon, chlorides, iron, and foaming to help pinpoint the problem.

Suspended Solids

Considered to be those solids and tarry hydrocarbons that remain suspended within the glycol solution down to 0.45 micron in size.

Result of poor inlet separation, corrosion, and thermal degradation of the glycol.

Values greater than 0.01% by weight indicate poor sock/microfiber filtration.

Most filters are sized to remove particles to a size of 5 microns.

Particles larger than this in excessive amounts may serve to stabilize foaming tendencies in glycol.

When the glycol is allowed to maintain a large concentration of suspended solids, a silty residue is likely to form along vessel walls.

Plugging of the contractor trays, heat exchangers, still column, and reconcentrator glycol is likely (common with low glycol pH).

Residue

The value for residue is a function of system contamination.

The glycol sample is distilled, removing all light end hydrocarbons, water, and glycol.

Residue represents the remaining contamination, which is comprised of:

> Total solids (suspended and residual)
>
> Salt
>
> Heavy hydrocarbons

Value for residue is best kept below 2% by weight, however some systems may operate reasonably well at values from 2% to 4%.

Units with Glycol containing greater than 4% are prime candidates for failure and should be cleaned immediately.

Chlorides

Chloride values indicate the quantity of inorganic chlorides (salts) found in the glycol sample.

As the concentration of chlorides (as NaCl or CaCl) in glycol increases, its solubility decreases.

Solubility also decreases when heat is added to the glycol solution.

When the solubility decreases, the salt begins to form crystals which:

> Fall out of the glycol solution
>
> Accumulate on the heat source and can lead to premature heat tube failure
>
> May be swept by the glycol into other areas of the system

Potential problems with excessive chlorides include system plugging, low pH, glycol pump damage, foaming, and loss of hygroscopicity due to rapid glycol decomposition.

Removal of chlorides in high concentrations requires vacuum distillation of the glycol.

Concentrations greater than 1000 ppm will stabilize foaming tendencies.

May lead to excessive glycol loss

May affect glycol pH

Precipitation of salts from the glycol will begin at approximately 12,200 to 1500 ppm, however, the crystals formed are extremely small and rarely troublesome.

Concentrations above 2200 ppm precipitation occurs readily and system failure is a possibility.

Filtration removes large salt crystals, but most of the damage associated with salt will have already occurred prior to the development of crystals large enough to filter.

Iron

Iron found in glycol samples can indicate:

Possible corrosion

Produced water carry-over

Iron in excess of 50 ppm generally indicates corrosion.

Whether it be in the glycol unit, upstream in the production equipment or downhole in the well string is difficult to determine.

Comparing values for pH, chlorides, and visual inspection of the glycol unit may help to establish the location of suspected corrosion.

Corrosion by-products will consist of soluble iron and fine, gritty particulate in systems where oxygen is available.

In systems where no oxygen is present, corrosion by-products will include sulfides in addition to the iron.

Foaming
General Considerations

More glycol is lost through foaming than any other cause.

Not easily detected without chemical analysis, gradual low-volume glycol loss often goes overlooked.

It is almost always a result of contamination.

Primary contaminants that cause foaming are:

Hydrocarbons (from separator carry-over) and solids

Chlorides, compressor lube oil, well-treating chemicals, and iron

Water content affect foaming tendencies by inducing emulsification of contaminants, particularly hydrocarbons.

Carbon filtration is the most effective means of controlling foam.

Silicone emulsion–type foam inhibitor is used, but they treat the symptom, not the cause and thus are temporary solutions.

Addressing the source of the contamination causing the foam is the only long-term solution.

Foam Test

The foam test consists of bubbling dry air at a rate of 6 liters/min through a graduated cylinder container of 200 mm of the glycol sample until the foam stabilizes at its maximum height.

Volume for both the liquid and the foam is reported as a single value.

The original 200 ml is then subtracted.

The remaining value is recorded as height and represents the ease at which the solution will foam.

Once the maximum foam height is recorded, the dry air is removed from the sample and the time it takes for the foam to break from its maximum volume to a clear surface on the glycol sample is recorded in seconds.

This time represents the tendency of the foam and is known as stability.

There are no concrete values given for acceptable foam height and stability.

Foam with very low height and moderate stability will result in little glycol loss as will a foam with moderate height and very low stability.

Thus, the acceptable range for foam test results are:

Height/ml: 20 to 30 ml

Stability/sec: 15 to 5 sec

Acceptable Limits

> Represents the increasing and decreasing values of acceptability for each.

> For example, a sample with a height of 25 ml and a stability of 10 sec is acceptable, while a sample with 30 ml height and 15 sec stability would have a high foaming tendency and could result in glycol losses.

3

Specific Gravity

Used to determine the purity of glycol.

A specific gravity of 1.126 to 1.128 at 60°F indicates a 99% TEG (technical grade).

A specific gravity of 1.124 to 1.126 indicates 97% (industrial grade).

With glycol extracted from an operating dehydration unit, the lean sample should have a specific gravity of 1.1189 to 1.121.

This variance allows for acceptable amounts of system contamination.

Low specific gravity would indicate one or more of the following:

> TEG containing excessive amounts of EG and/or DEG (poor quality replacement glycol)

> Excessive water in sample

> Excessive hydrocarbons in sample

A high specific gravity indicates the system is contaminated with excessive amounts of solids or any additives with a greater density than glycol:

> Thermal degradation of the glycol

> Oxidation or chemical degradation of the glycol

Glycol Composition

The composition of glycol indicates its quality.

Values are given to the component glycols (EG, DEG, TEG, TTEG) contained within the glycol sample solution.

Industrial grade (97%) TEG or better is required for best glycol system results.

In addition to 97% TEG, the glycol solution may contain, in various concentrations, up to 1% EG and 3% DEG, but not to exceed a combined total of 3%.

Glycol degradation will often be reflected by changes in the glycol composition and reduction in pH.

Thermal degradation is most common and is characterized by excessive values of EG, DEG, and occasionally the presence of TTEG.

The glycol pH will be low.

The glycol sample will be dark and have an aromatic smell (ripe bananas).

Chemical degradation is brought about by oxidation and acidic contaminants and is characterized by:

> Excessive values for EG and DEG but no TTEG will be present

> Low pH

> Glycol may not appear to be too dirty

Autoxidation is a form of continuing chemical degradation.

Troubleshooting

General Considerations

Even the best preventive maintenance program will not guarantee that the dehydration unit will operate trouble-free.

The most obvious indication of a unit malfunction is high water content (dew point) of the outlet stream.

High water content is brought about by:

> Insufficient glycol circulation

> Reconcentration of the glycol

These problems can be caused by a variety of contributing factors such as:

> Mechanical causes

> Existing operating conditions for which the equipment was not designed

These conditions can sometimes be at least partially alleviated by both changes in condition and mechanical operation.

High Dew Points
Insufficient Glycol Circulation

If there is insufficient glycol circulation, check heat exchangers and glycol piping for restrictions or plugging.

On an electric driven piston pump:

> Check flow rate indicator (if present) to insure proper glycol circulation. If flow rate indicator is not present, verify circulation rate by closing the glycol discharge valve from the contactor and timing the fill rate in the gauge column.
>
> Check high-pressure dry-glycol bypass valve. Close if necessary.
>
> Check pump prime by shutting pump down, closing the discharge valve, opening the bypass valve and restarting the pump. Allow to run briefly under no load through the bypass line to remove any trapped gas in the pump.

On glycol-gas powered pumps:

> Close dry discharge valve. If pump continues to run, open dry discharge bleed valve and allow to run a few strokes. Once all gas is purged from put, close the bleed valve. If pump continues to run, discontinue use and send in for repair.

If pump will not prime, but continues to run gas through the dry discharge bleed valve then:

> Check pump suction strainer for plugging.
>
> Check glycol level in surge tank.

Insufficient Reconcentration

Verify reconcentration temperature with test thermometer (350° to 400°F). Raise temperature if necessary.

Check glycol-to-glycol heat exchanger for leakage of wet glycol into the dry glycol stream.

Check stripping gas if applicable. Be sure stripping gas is in service at the proper rate.

Check for communication between the reconcentration vapor space and the surge tank vapor space. Communication could mean contaminated dry glycol going to the pump.

3

Operating Conditions Different from Design

Check operation of upstream separators and scrubbers. Be sure not to overload system.

Increase absorber pressure. This may require installation of a back pressure valve.

Reduce gas temperature, if possible.

Increase circulation rate, if possible.

Increase reconcentrator temperature, if possible.

Low Flow Rate

Blank off a portion of the bubble caps, if possible.

Lower system pressure.

Add additional cooling to dry glycol and increase circulation rate.

Change out absorber to a small unit designed for a lower flow rate.

Absorber Tray Damage

Open inspection ports and/or manway and verify tray integrity. Repair or replace as necessary.

Breakdown or Contamination of Glycol

Have lean and rich glycol sample analyzed. Note evidence of severe contamination, thermal or chemical decomposition. Clean system and/or recharge with fresh glycol as necessary.

Glycol Loss from the Contactor

Foaming

Major cause of foaming is contamination. Remove source of contamination. Clean contactor if necessary, clean system if necessary, replace glycol if necessary.

Increase filter capacity and/or add carbon filtration.

Add antifoam compound (silicon emulsion type).

Adjust high pH to prevent emulsification (use acetic acid).

Plugged or Dirty Trays

Clean trays.

Manually enter tower and clean.

Open inspection ports and clean with water jet or by hand.

Chemically clean.

Excessive Velocity

Decrease gas rate.

Increase absorber pressure.

Interrupted Liquid Seal on the Trays (Gas Surge)

If the contactor has a bypass valve, isolate the tower by opening the bypass valve and closing the gas inlet valve. Allow the glycol pump to run 5 minutes then while the glycol is circulating open the gas inlet valve and slowly close the gas bypass valve.

If contactor does not have a bypass valve, stop or greatly reduce the gas flow through the tower (shut wells, flare gas, alternate system, etc.). Allow the glycol to circulate 5 minutes then slowly turn the gas back through the tower.

If unable to stop or reduce the gas flow, increase the glycol circulation rate to the maximum possible for 2–5 minutes (flood trays in attempt to reestablish seal using liquid head pressure).

Cold Glycol (Cold Gas)

Increase gas temperature by increasing temperature of flowline heater or add flowline heater, if necessary.

Leaks

Pressure test external gas glycol heat exchanger for glycol leakage into dry gas stream.

Check drain header (if applicable) at all gauge columns, external float cages (LSLL, etc.).

Accumulation in Integral Scrubber

Check for communication between chimney tray and scrubber section.

Check bottom tray leakage. May have damaged or improperly constructed chimney.

Check glycol level control and dump valve operation (on units with electric powered glycol pumps).

Glycol Loss from the Reconcentrator

Leaks

Be sure all drain valves are closed.

Be sure gauge column seals are good.

Check heat tube integrity (glycol loss into fire tube or waste heat tube will produce heavy smoke from stack).

Check reconcentrator shell integrity (note glycol leakage from insulation, wet insulation, or telltale stains).

Heat source flange leak (poor gasketing).

Bad Glycol Relief Valve

Replace glycol relief valve.

Exiting the Still Column

For plugged or fouled still column packing, clean or replace still column packing.

For saturated glycol (droplets blowing out still):

Check reconcentrator heat source. Insure heat is between 350° and 400°F.

Check for free liquid or misting liquid carry-over into contactor tower. Repair or replace separator control, if necessary. Reduce slugging if possible. Add scrubber, if necessary.

Reduce glycol flow through the reflux condenser. Raise reflux temperature.

Vaporization

Check reconcentrator temperature (below 404°F).

Check reflux temperature. Increase the glycol flow through the reflux condenser to lower the reflux temperature.

Check stripping gas flow rate.

Check for plugged or fouled glycol outlet from reconcentrator (downcomer or heat exchangers).

Glycol Loss—Glycol Hydrocarbon Separator

Improper Control Operation

Repair or replace level control.

Clean, repair, or replace dump valve.

Leaks

Check drain valve. Tighten, repair, or replace.

Check gauge columns, external float cages, and level control adapters.

Add antifoam compound to prevent loss through gas outlet.

Accumulation in Oil Bucket (Bucket-and-Weir)

Open vessel and clean glycol passage under oil bucket (horizontal vessels).

Adjust or remove weir.

Glycol Loss—Miscellaneous

Leaks

Check all flanges, unions and associated piping.

Check electric pump rod packing.

Check all drain valves (filter, heat exchanger, etc.).

Check pump bleed valves (and electric pump bypass).

Check external gas-glycol heat exchanger.

Poor Quality or Contamination Replacement Glycol

Use only virgin TEG with 97% or better purity.

Check glycol content for excessive water.

Three-Step Approach to Troubleshooting

Timeframe

Determine the approximate date/time at which the problem became apparent.

List Changes

Inventory any changes (things that happened differently than usual). Look for what is different.

Production changes

Operational changes

Maintenance

Repairs

Weather

Investigate

By process of elimination reduce the list of changes to determine the factor or factors that manifest the problem.

Glycol System Cleaning

General Considerations

Chemicals are frequently needed to clean the glycol system.

> If chemical cleaning is done properly, it can be quite beneficial to plant operations.

> If done poorly, it can be quite costly and create long-lasting problems.

The most effective type of cleaner is a very heavy-duty alkaline solution.

To provide optimum cleaning, the concentration, temperature, and pumping rate of the solution must be carefully controlled and an experienced, reputable vendor employed.

A cascading technique can be used to save on the cost of cleaning chemicals.

Cleaning Techniques to Avoid

Steam cleaning is not effective and can be damaging and dangerous. It tends to harden the deposits in the system, making them almost impossible to remove.

The use of cold or hot water, with or without high detergent soaps, will do little good in cleaning the system.

> High-detergent soaps can create a serious problem by leaving trace quantities of soap after the cleaning job.

> Soap traces left in the system can make glycol foam for a long time.

Acid cleaning is good for removing inorganic deposits

> Since most deposits in the glycol system are organic, acid cleaning is not very effective.

> It can easily create additional problems in the glycol system after the cleaning job.

ELIMINATING OPERATING PROBLEMS

General Considerations

Most operating problems are caused by mechanical failure.

It is important to keep equipment in good working order.

Following operating and maintenance suggestions helps provide a trouble-free operation.

Inlet Scrubber/Microfiber Filter Separator

The cleaner the inlet gas entering the absorber, the fewer operating problems there will be.

Potential problems if there wasn't an inlet scrubber of filter separator are:

Liquid water carryover

>> Dilutes the glycol

>> Lowers the absorber efficiency

>> Requires a greater glycol circulation rate

>> Increases the vapor-liquid load on the still column

>> Floods the still column

>> Vastly increases the reboiler heat load and fuel gas requirements

If the water contained salt and solids, they would be deposited in the reboiler to foul the heating surfaces and possibly cause them to burn out (Figures 3-12 through 3-14)

If liquid hydrocarbons were present:

>> They would pass onto the still column and reboiler

>> Lighter fractions would pass overhead as vapor and could create a fire hazard

>> Heavy fractions would collect on the glycol surface in the storage tank and could overflow the system (Figure 3-15)

Flashing of the hydrocarbon vapor can flood the still column and vastly increase the reboiler heat load and result in glycol losses

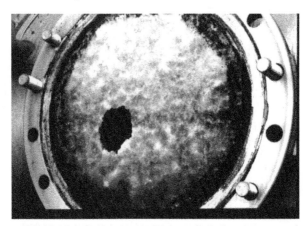

FIGURE 3-12 Salt deposited in interior of reboiler. (Courtsey of Gly-Tech)

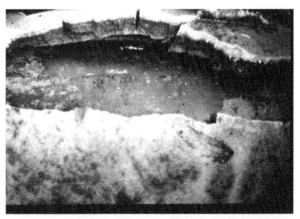

FIGURE 3-13 Salt fouling fire tube. (Courtsey of Gly-Tech)

Well corrosion control program should be planned and coordinated to prevent glycol contamination:

Excessive fluid will carry over into the system if the inlet scrubber/filter separator is overloaded.

3

FIGURE 3-14 Salt covering fire tube perimeter. (Courtsey of Gly-Tech)

FIGURE 3-15 Heavy fractions of tarry hydrocarbons collected from bottom of reboiler.

Gas from the treated wells should be slowly passed through a tank or separate system at the wellhead until the corrosion inhibitor and distillate carrier can be collected.

Do not open all the treated wells at one time. This will keep large slugs out of the gathering lines going to the plant.

Scrubber or filter separator may be an integral part of the absorber or preferably a separate vessel

> Vessel should be large enough to remove all solids and free liquids to keep these impurities from getting into the glycol system

> Vessel should be regularly inspected to prevent any malfunction

Liquid dump line should be protected from freezing during cold weather:

> Accomplished with a heating coil in the scrubber or separator

> Warm glycol is pumped through the coil

> Flow is directed through the coil by means of block and bypass valves

> Separator may be provided with a heating chamber on the liquid level controller and in the gage glass

> Cold-weather provision may include a heating coil in the reboiler to heat purge gas which can be bled into the separator liquid dump line to keep the liquid moving so it does not freeze

Separator should be located close to the absorber so the gas does not condense more liquids before it enters the absorber

If a separator ahead of the glycol plant is equipped with a pressure safety valve, a flow safety valve should be installed at the inlet to the absorber to protect the internals

Sometimes an efficient mist extractor, which removes all contaminants over one micron is needed between the inlet separator and the glycol plant to clean the incoming gas; this is particularly useful when paraffin and other impurities are present in a fine vapor form

When gas is compressed prior to dehydration:

> A coalescing type of scrubber (microfiber filter separator) placed ahead of the absorber insures removal of compressor oil in vapor form.

> Compressor oil and distillate can coat the tower packing either in the absorber or still column and decrease its effectiveness.

Absorber

This vessel contains valve or bubble cap trays or packing to give good gas-liquid contact.

Cleanliness is very important to prevent high sales gas dew points caused by foaming and/or poor gas-liquid contact.

Plugged trays or packing could also increase glycol losses.

Unit startup considerations are as follows:

The pressure on the absorber should be slowly brought up to the operating range and then the glycol should be circulated to get a liquid level on all trays.

Next, the gas rate going to the absorber should be slowly increased until the operating level is reached.

If the gas enters the absorber before the trays are sealed with liquid, it will pass through the downcomers and bubble caps.

When this condition occurs and the glycol is pumped into the absorber, the liquid has difficulty in sealing the downcomers.

Liquid will be carried out with the gas stream instead of flowing to the bottom of the absorber.

Gas flow rate should be increased slowly when changing from a low to a high flow rate.

Rapid surges of gas through the absorber may cause:

Sufficient pressure drop through the trays to break the liquid seals, and/or

Glycol to be lifted off the trays, which will flood the mist extractor and increase glycol losses

Unit shutdown considerations are as follows:

First, the fuel to the reboiler should be shut down.

Then the circulating pump should be run until the reboiler temperature is lowered to approximately 200°F (94°C).

This precaution will prevent glycol decomposition caused by overheating.

3

The unit can then be shut down by slowly reducing the gas flow to prevent any unnecessary shocks on the absorber and piping.

The unit should be depressurized slowly to prevent a loss of glycol.

The dehydrator should always be depressurized from the downstream (gas outlet) side of the absorber.

A dehydrator installed on the discharge side of a compressor should be equipped with a check valve in the inlet line, located as close as possible to the absorber.

Experience has shown that some glycol is sucked back into this line when a compressor backfires or is shut down.

Internal absorber damage to the trays and mesh pad may also occur with a compressor failure. (Figure 3-16).

The installation of the check valve usually eliminates this problem.

All compressors taking gas from or feeding gas to a dehydrator should have pulsation dampeners.

The absence of this safety device may cause fatigue failure of instruments, trays, coils, mesh pads and other parts of the dehydrator.

FIGURE 3-16 Tray damage due to compressor failure.

The glycol dump valve and level controller should be set for throttling action to give an even flow of glycol to the regenerator.

> This will prevent slugs, which could flood the stripper and cause excessive glycol losses.

The absorber must be vertical.

> Insure the proper flow of glycol in the vessel and adequate contact of the glycol and gas.

> Sometimes the trays and bubble caps do not seal properly after erection and should be inspected if very high glycol losses exist.

> Inspection ports at the trays can be very useful when inspecting or cleaning the vessel.

If dry gas from a glycol unit is used for gas lift:

> Care must be used in both sizing and operating the unit because of the unsteady gas rate required in this service.

> A back pressure valve should be installed on the gas outlet from the absorber operating on a gas lift system.

>> If this is not done, then a valve downstream of the absorber can be pinched to prevent a sudden overloading of the absorber and helps control the gas flow through the unit.

>> A sudden overloading of the absorber can break the downcomer seals in a tray type of vessel and cause excessive loss of glycol in the sales gas.

Absorbers sometimes need to be insulated when excessive condensation of light hydrocarbons collect on the vessel walls.

> This often occurs when dehydrating rich, warm gases in cold climates.

> These very light hydrocarbons can cause tray flooding in the absorber and excessive glycol losses from the regenerator.

The mist extractor should receive special attention because glycol entrainment and well-crawling are difficult to effectively control. (See Figures 3-17 through 3-20.)

FIGURE 3-17 Partially plugged mist extractor.

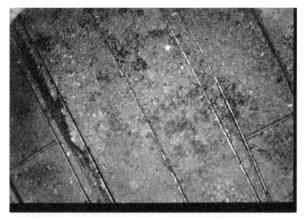

FIGURE 3-18 Completely plugged mist extractor.

The type and thickness of the mesh pad should be carefully studied to minimize glycol losses.

Care should also be taken after installation to avoid mesh pad damage.

The maximum pressure drop through the contractor to avoid damage to the mesh pad is approximately 15 psi.

FIGURE 3-19 Replacement mist extractor. (Courtsey of Gly-Tech)

FIGURE 3-20 Installation of replacement mist extractor.

Glycol-Gas Heat Exchanger

Most units are supplied with a glycol-gas heat exchanger that uses the gas leaving the absorber to cool the lean glycol entering the absorber.

> This exchanger may be a coil in the top of the absorber or an external one.
>
> A water-cooled exchanger may be used when heating of the gas must be avoided.

This exchanger may accumulate deposits, such as salt, solids, coke or gum which:

Foul the heat exchanger surface

Reduce the heat transfer rate

Increase the lean glycol temperature

All of the above increase glycol losses and make dehydration difficult.

The vessel should be inspected regularly and cleaned when needed.

Lean Glycol Storage Tank or Accumulator

Normally this vessel contains a glycol heat exchanger coil which does the following:

Cools the lean glycol coming from the reboiler

Preheats the rich glycol going to the stripper

The lean glycol is also cooled by radiation from the shell of the storage tank.

This accumulator should normally be insulated.

Water cooling can also be used to help control the lean glycol temperature.

On conventional regenerators without stripping gas:

Accumulator must be vented to prevent trapping gas

Vapors, trapped in the storage tank, could cause the pump to vapor lock

A connection is usually provided in the top of the storage tank for venting

Vent line should be piped away from the process equipment but should not be connected to the stripper vent because this could cause steam to dilute the concentrated glycol

Some units are equipped to provide a dry gas blanket (no oxygen or air) in the storage tank.

It's usually not necessary to hook up a separate vent on these storage tanks.

Blanket gas is normally piped to the regular vent connection on top of the storage tank.

If blanket gas is used, it is commonly taken from the fuel gas line.

When blanket gas is used, it may be necessary to see that the blanket gas valve, piping, and flow control orifice are open to pass gas.

Only a very slight flow of gas is required to prevent steam generated in the reboiler from contaminating the regenerated glycol.

The vessel should be inspected occasionally to see that sludge deposits and heavy hydrocarbons are not collecting in the bottom of the vessel.

The heat exchanger coil should be kept clean so proper heat transfer can be made. This also prevents corrosion.

If the heat exchanger develops a leak, the water rich glycol could dilute the lean glycol.

Glycol level in the storage tank should be checked and a level in the gauge glass should always be maintained.

Gauge glass should be kept clean to assure an optimum level.

Glycol should be added as the level is pumped down.

Records of the amount of glycol added should be maintained.

Make certain the storage tank is not overfilled as this could present problems as well.

Stripper or Still Column

The stripper, or still column, is generally a packed column located on top of the reboiler to separate the water and glycol by fractional distillation.

Packing is usually a ceramic saddle but 304 stainless steel pall rings can be used to prevent breakage. (See Figures 3-21 through 3-24.)

A standard stripper usually has a finned atmospheric condenser in the top to cool the steam vapors and recover the entrained glycol.

Atmospheric condenser depends upon air circulation to cool the hot vapors.

Increased glycol losses can occur on extremely hot days when insufficient cooling in the condenser causes poor condensation.

FIGURE 3-21 Ceramic saddle packing.

FIGURE 3-22 Stainless steel pall ring packing.

High glycol losses can also occur on extremely cold, windy days when excessive condensation (water and glycol) overloads the reboiler.

Excess liquids percolate out the stripper vent.

FIGURE 3-23 Structured packing.

FIGURE 3-24 Ceramic saddles coated with hydrocarbon tar.

If stripping gas is used, an internal reflux coil is normally provided to cool the vapors.

Reflux for the stripper is more critical when stripping gas is used to prevent excessive glycol losses.

This is due to a larger mass of vapor leaving the stripper which will carry glycol.

Adequate reflux is provided by passing the cool, rich glycol from the absorber through the condenser coil in the stripper.

If properly adjusted, it can provide uniform condensation throughout the year.

Manual/automatic valve in the piping is furnished to bypass the reflux coil.

Under normal circumstances this valve will be closed and the total flow will be through the reflux coil.

In cold weather operation, with extreme low ambient temperatures, this could produce too much reflux and the regenerator could become overloaded.

The reboiler may not be able to maintain the required temperature.

With these conditions, the ambient air is providing part or all of the reflux required.

Therefore, a portion or all of the rich glycol solution should bypass the reflux coil.

This is accomplished by opening the manual/automatic valve until the reboiler can hold the temperature.

This lowers the amount of reflux produced by the coil and reduces the load on the reboiler.

Sometimes a leak can develop in the cool glycol reflux coil in the top of the stripper. When this happens, excess glycol can:

Flood the tower packing in the still column

Upset the distillation operation

Increase glycol losses

The reflux coil should be properly maintained.

Broken, powdered packing can cause solution foaming in the stripper and increase glycol losses.

Packing is usually broken by excessive bed movement which is caused when hydrocarbons flash in the reboiler.

Careless handling when installing the packing can also cause powdering.

As particles break down, the pressure drop through the stripper increases.

This restricts the flow of vapor and liquid and causes the glycol to percolate out the top of the stripper.

Dirty packing, caused by sludge deposits of salt or tarry hydrocarbons, will also cause solution foaming in the stripper and increase glycol losses.

Packing should be cleaned or replaced when plugging or powdering occurs.

The same size tower packing should be used for replacement.

The standard size of the ceramic saddle or a stainless steel pall ring is one inch.

When stripping gas is used and a tower packing is placed in the downcomer between the reboiler and the storage tank, provisions should be made to replace the tower packing without cutting into the downcomer.

During low circulation rates:

Rich glycol may channel through the packing, causing poor contact between the liquid and hot vapors.

To prevent channeling, a distributor plate can be placed below the rich glycol feed line to evenly spread the liquid.

A large carryover of liquid hydrocarbons into the glycol system can be very troublesome and dangerous.

The hydrocarbons will flash in the reboiler, flood the stripper, and increase glycol losses.

Heavy hydrocarbon vapors and/or liquids could also spill over the reboiler and create a serious fire hazard.

Therefore, the vapors leaving the stripper vent should be piped away from the process equipment as a safety measure.

The vent line should be properly sloped all the way from the stripper to the point of discharge to prevent condensed liquids from plugging the line.

If the vent line is long and is carried above the ground, a top vent, at a point not over 20 feet away from the stripper, should probably be installed to allow the escape of vapors in case of a freeze-up in the long line.

The piping should be the same size as the vessel connection, or larger.

In areas where there is a possibility of cold, freezing weather, the vent line should be insulated from the stripper to the discharge point to prevent freeze-ups.

This will prevent the steam from condensing, freezing, and plugging the line.

If freezing occurs, the water vapor flashed in the reboiler may discharge into the storage tank and dilute the lean glycol.

The pressure caused by these trapped vapors could also force the regenerator to burst.

Reboiler

The reboiler supplies heat to separate the glycol and water by simple distillation.

Large plant locations may use hot oil or steam in the reboiler.

Remote field locations are generally equipped with a direct-fired heaters (Figure 3-25), with the following characteristics:

Use a portion of the gas for fuel

Heating element usually has a U-Tube shape and contains one or more burners

Should be conservatively designed to:

Insure long tube life

Prevent glycol decomposition caused by overheating (see Figures 3-26 and 3-27)

Reboiler should be equipped with a high-temperature safety overriding controller to shut down the fuel supply gas system in case the primary temperature controller should malfunction

The firebox heat flux (a measure of the heat transfer rate in Btu/hr/ft^2) issues include:

It should be high enough to provide adequate heating capacity but low enough to prevent glycol decomposition.

Excessive heat flux, a result of too much heat in a small area, will thermally decompose the glycol (see Figure 3-28).

FIGURE 3-25 Direct-fired reboiler.

FIGURE 3-26 Decomposed glycol on reboiler fire tube.

The pilot flame should be kept low, especially in small reboilers for the following reasons:

Prevents glycol decomposition

Prevents tube burnout (see Figure 3-29)

Particularly important on smaller units where the pilot flame can supply a substantial portion of the total heat requirement

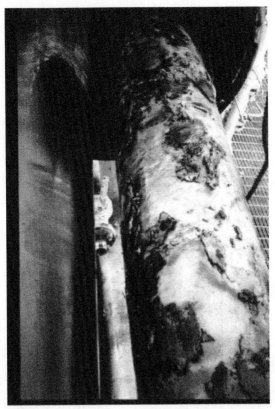

FIGURE 3-27 Decomposed glycol on reboiler fire tube. (Courtsey of Gly-Tech)

Flame should be correctly adjusted to give a long, rolling, and slightly yellow-tipped flame

Nozzles are available that distribute the flame more evenly along the tube:

Decreases the heat flux of the area nearest the nozzle without actually lowering the total heat energy transferred

Avoids direct and hard impingement of the flame against the firetube

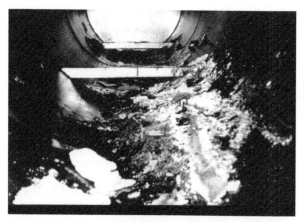

FIGURE 3-28 Decomposed glycol in reboiler vessel shell.

FIGURE 3-29 Decomposed glycol with fire-tube burnout.

A pump shutdown device can prevent the circulation of wet glycol of wet glycol caused by a flame failure

A continuous spark ignition system, or a spark igniter to relight the pilot if it goes out, is also useful

Orifices on the air-gas mixers and pilots should be cleaned regularly to prevent burner failures

The following temperatures in the reboiler should not be exceeded to prevent burner failures:

Type of Glycol	Thermal Decomposition Temperatures
Ethylene	329°F (165°C)
Diethylene	328°F (164°C)
Triethylene	404°F (207°C)

Excessive discoloration and very slow degradation will result when the reboiler bulk temperature is maintained about 10°F (5°C) in excess of the above listed temperatures.

If coke, tarry products, and/or salt deposit on the firetube, the heat transfer rate is reduced and a tube failure can result.

Localized overheating, especially where salt accumulates, will decompose the glycol.

An analysis of the glycol determines the amounts and types of these contaminants.

Salt deposits can also be detected by shutting off the burner on the reboiler at night and looking down the firebox.

A bright red-glowing light will be visible at spots on the tubes where salt deposits have collected.

These deposits can cause a rapid firetube burnout, particularly if the plant inlet separator is inadequate and a slug of salt water enters the absorber.

Coke and tarry products present in the circulating glycol can be removed by good filtration.

More elaborate equipment is needed to remove the salt.

Contaminates, which have already deposited on the firetube and other equipment, can only be removed by using chemicals.

The heating process is thermostatically controlled and fully automatic.

The reboiler temperature should be occasionally verified with a test thermometer to make sure true readings are being recorded.

If the temperature fluctuates excessively when operating below the design capacity, the fuel gas pressure should be reduced.

A uniform temperature gives a better operation of the reboiler.

If the reboiler temperature cannot be raised as desired, it may be necessary to increase the fuel gas pressure up to about 30 psig.

If water and/or hydrocarbons enter the reboiler from the absorber, it may be impossible to raise the temperature until this problem is corrected.

Standard orifices furnished for reboiler burners are sized for 1000–1100 Btu/scf of gas.

If the rating of the fuel gas is less than this, it may be necessary to install a larger orifice or drill out the existing orifice to the next higher size.

Fires have been caused by leaks in the gas lines near the firebox.

The best precaution is to place valves and regulators in the gas line at a maximum distance from the firebox.

Another effective measure is the addition of a flame arrestor around the firebox.

If the arrestor is properly designed, even severe gas leaks in the immediate vicinity of the firebox will not ignite.

During a unit startup, it is imperative the reboiler temperature be up to the desired operating level before flowing gas through the absorber.

The reboiler must be horizontal when erected.

A nonhorizontal position can cause a firetube burnout.

The reboiler should also be located close enough to the absorber to prevent excessive cooling of the lean glycol during cold weather.

This will prevent hydrocarbon condensation and high glycol losses in the absorber.

Stripping Gas

Stripping gas is an optional item used to achieve very high glycol concentrations which cannot be obtained with normal regeneration.

It will provide the maximum dew point depression and greater dehydration.

Stripping gas is used to remove the residual water after the glycol has been reconcentrated in the regeneration equipment.

It is used to provide intimate contact between the hot gas and the lean glycol after most of the water has been removed by distillation.

Lean glycol concentrations in the range of 99.5–99.9% and dew-point depressions of 140°F and above have been reported.

There are several methods of introducing stripping gas into the system.

One method is to use a vertical tray or packed section in the downcomer between the reboiler and storage tank where the dry gas strips the additional water out of the regenerated glycol.

The glycol from the reboiler flows down through this section, contacts the stripping gas to remove the excess water, and goes into the storage tank.

Another method is to use glycol stripping gas sparger in the reboiler beneath the firetube.

As the glycol flows through the reboiler, gas is injected into this vessel and is heated by the glycol.

Stripping gas contacts the glycol in the reboiler and removes some of the additional water.

Gas then passes out the stripper to the waste pit.

The lean glycol flows from the reboiler down into the storage tank.

Stripper gas is normally taken from the reboiler fuel gas line (if dehydrated gas) at the fuel drip pot pressure.

Air or oxygen should not be used.

Stripping gas is usually controlled by a manual valve with a pressure gauge to indicate the flow rate through an orifice.

Stripping gas rate has the following characteristics:

Will vary according to the lean concentration desired and the method of glycol-gas contact

Usually between 2 and 10 scf/gallon of glycol circulated

Should not get high enough to flood the stripper and blow glycol out to the pit

When stripping gas is used it is necessary to provide for more reflux in the stripper to prevent excessive glycol losses.

This is usually provided by using a cool glycol condenser coil in the stripper.

Circulating Pump

A circulating pump is used to move glycol through the system.

It can be powered by electricity, gas, steam, or gas and glycol, depending upon the operating conditions and unit location.

A gas-glycol pump, a very versatile piece of equipment, is commonly used for the following reasons:

Controls are serviceable, dependable, and, if adjusted properly, should give a long, trouble free operation

Utilizes the rich glycol under pressure in the absorber to furnish part of its required driving energy

Since the pump cannot get more glycol back than it pumped over, a supplemental volume is needed to provide the driving force

Gas, under pressure from the absorber, is taken with the rich glycol to supply this additional volume

At 1000 psi operating pressure on the absorber, the volume of gas required is approximately 5.5 scf per gallon of lean glycol circulated

Helpful maintenance tips:

Careful starting of a new pump can save much worry and downtime.

Pump packing gland generally used is lubricated only by the glycol itself.

Packing is dry when the pump is new.

As it soaks up glycol, the packing tends to expand.

If it has been screwed down too tight, either the packing will score the plunger or the packing will burn out.

The pump normally handles a fluid that is frequently dirty and corrosive.

Can lead to cylinder corrosion, seal erosion, impeller damage, pump cup or ring wear, and sticking or plugged valves.

These parts must be checked and kept in proper condition to keep the pump at maximum efficiency.

Pump rate should be commensurate with the gas volume being processed.

Speed should be decreased for low gas rates and increased for high rates.

Proportioning adjustments allow increased gas-glycol contact time in the absorber.

When the pump check valves become worn or clogged, the pump will operate normally except no fluid will go to the absorber.

Even a pressure gauge will indicate a pumping cycle.

The only evidence of this type of failure is little or no dew-point depression.

One sure way to check the volume flowing is to close the valve on the absorber outlet and calculate the flow by measuring the rise in the gauge glass (if one is available) versus the amount pumped normally

One of the most common sources of glycol loss occurs at the pump packing gland.

> If the pump leaks over one or two quarts of glycol per day the packing needs to be replaced.

> An adjustment will not recover the seal.

> Packing should be installed hand-tight and then backed off one complete turn.

> If the packing gets too tight, the pistons can score and require replacement.

Glycol circulation rate of 2 to 3 gallons/lb of water to be removed is sufficient to provide adequate dehydration.

> An excessive rate can overload the reboiler and reduce the dehydration efficiency.

> The rate should be checked regularly by timing the pump to make sure it is running at the proper speed.

Proper pump maintenance will reduce the operating costs.

> When the pump is not working the whole system must be shut down because the gas cannot be dried effectively without a good continuous flow of glycol in the absorber.

> Therefore, small replacement parts should be readily available to prevent lengthy shutdowns.

If there is insufficient glycol circulation:

> Check the pump suction strainer for plugging and/or open the bleeder valve to eliminate air lock.

> Glycol strainers should be regularly cleaned to avoid pump wear and other problems.

Pumps should be lubricated regularly.

Easy access to the pump can save time and trouble when making repairs and replacing components.

The maximum operating temperature of the pump is limited by the moving O-ring seals and nylon D slides.

A maximum temperature of 200°F (94°C) is recommended.

Packing life will be extended considerably if the temperature is held to a maximum of 150°F (66°C).

Therefore, sufficient heat exchange is necessary to keep the dry, lean glycol below these temperatures when it goes through the pump.

The pump is usually the most overworked and overused piece of equipment in the glycol process system.

The glycol system usually contains a second spare pump to avoid shutdowns when the primary pump fails.

It is not uncommon for operators to use the second pump to send more glycol to the absorber to avoid wet sales gas problems.

This procedure just increases operating problems.

All of the other process variables should first be checked before a second pump is used.

A pressure gauge is furnished on the discharge side of the pump.

A valve is also furnished between the pressure gauge and the line so the pressure gauge can be isolated.

Pressure gauge can be used to see that the pump is working by watching the gauge "kick" as the pump piston strokes.

The sensing element in the pressure gauge is a bourdon tube.

The flexing or movement of this tube indicates the pressure.

A bourdon tube will fatigue or fail if subjected to continuous fluctuations in pressure on the pump discharge.

Pressure should be kept off the gauge except when testing the unit or to determine glycol loss from the gauge failure.

Flash Tank or Glycol-Gas Separator

The flash tank, or glycol-gas separator, is an optional piece of equipment used to recover the off-gas from the glycol-powered pump and the gaseous hydrocarbons from the rich glycol.

The recovered gas can be used as fuel to the reboiler and/or stripping gas.

Any excess gas is usually discharged through a back pressure valve.

The flash tank will keep volatile hydrocarbons out of the reboiler.

This low-pressure separator may be located in one of the following two places:

Between the pump and the preheat coil in the storage tank

Between the preheat coil and the stripper

The separator usually works best in a temperature range of 130°F to 170°F (55°C to 77°C).

A two-phase separator, with at least a five minute retention time, can be used to remove the gas.

If liquid hydrocarbons are present in the rich glycol, a three-phase separator should be used to remove these liquids before they get in the stripper and reboiler.

A liquid retention time of 20 to 45 minutes, depending on the type of hydrocarbons, API gravity, and the amount of foam, should be provided in the vessel.

Vessels should be located ahead of or behind the preheat coil in the storage tank, depending on the type of hydrocarbons present.

Gas Blanket

A gas blanket prevents air from contacting glycol in the reboiler and storage tanks.

A small amount of low-pressure gas is bled into the storage tank.

Gas is piped from the storage tank to the bottom of the stripper and it passes on overhead with the water vapor.

Elimination of air helps prevent glycol decomposition by slow oxidation.

The gas blanket equalizes the pressure between the reboiler and storage tank. The gas blanket also prevents the liquid seal from breaking down between these two vessels.

Reclaimer

The reclaimer purifies the glycol for further use by vacuum distillation.

Clean glycol is driven off and all the dirty sludge is left in the vessel and then washed to the sewer.

It is normally used only in very large glycol systems.

IMPROVING GLYCOL FILTRATION

General Considerations

Filters will do the following:

Extend the life of pumps

Prevent an accumulation of solids in the absorber (see Figures 3-30 through 3-34)

Prevent an accumulation of solids in the regeneration equipment

Solids that settle out on metal surfaces will frequently set up cell corrosion.

Filters remove the solids to also eliminate fouling, foaming, and plugging.

Filters should be designed to remove all solid particles 5 microns and larger.

They should be able to operate up to a differential pressure of 20–25 psi without loss of seal or channeling of flow.

FIGURE 3-30 Top view of replacement of microfiber filter elements.

An internal relief valve with a setting of about 25 psi and differential pressure gauges are very helpful.

New elements should be installed before the relief valve opens.

For filters equipped with block and bypass valves:

The bypass valve should be opened first before the block valves are closed to prevent excess pressure on the unit.

FIGURE 3-31 Microfiber filters contaminated with liquid hydrocarbon carry-over.

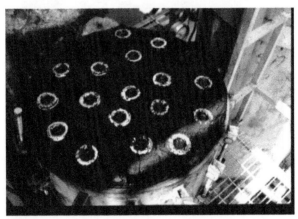

FIGURE 3-32 Decomposed glycol captured in the microfiber filters.

For filters not equipped with block and bypass valves:

Close the block valve on the absorber glycol dump line before attempting to change elements.

Filters are not usually placed in the rich glycol line for best results, but the lean glycol can also be filtered to help keep the glycol clean.

FIGURE 3-33 Decomposed glycol captured in the microfiber filters.

FIGURE 3-34 Filter collapsed due to high-pressure drop.

Frequent filter changes may be needed during unit startup or when neutralizers are added, to control the glycol pH.

New elements should be placed in a dry, clean place to keep them from dirt and grease.

Consult the filter manufacturer for installation and operating instructions.

It is important to know when and how to change elements to keep air out the glycol system.

Valves and gauges should be inspected occasionally for corrosion and scale buildup.

To determine the proper use of filter elements, cut them to the core and inspect them.

If they are dirty throughout, the filter is being used properly.

If the element is clean on the inside, an element with a different micron size may be needed.

It is also a good practice to occasionally scrape some sludge from a dirty element and have it analyzed.

This will help establish the types of contaminants present.

A record of the number of elements replaced will establish the amount of contaminants present.

USE OF CARBON PURIFICATION

General Considerations

Activated carbon can effectively eliminate most foaming problems by removing the hydrocarbons, well-treating chemicals, compressor oils, and other troublesome impurities from the glycol.

Two ways glycol purification can be achieved are as follows:

One method is to use two carbon towers installed in series but piped so they can be taken off-stream or interchanged without difficulty.

In large systems about 2% of the total glycol flow should pass through the carbon towers.

In small systems 100% of the total glycol flow should pass through the carbon towers.

Each carbon bed should be sized to handle 2 gallons of glycol per square foot of cross-sectional area per minute.

Towers should have an L/D ratio of about 3:1 to 5:1 and even 10:1 in some cases.

Towers should be designed to permit back-flushing with water to remove the dust after the carbon is loaded.

To achieve this, a retainer screen, with a smaller mesh size than the carbon should be installed above the carbon bed between the liquid inlet distributor and the outlet water drain nozzle to hold the carbon to the vessel.

The liquid distributor is needed to avoid glycol channeling through the carbon.

3

The screen size and support for the bottom of the towers should be carefully selected to avoid carbon plugging and to keep the carbon in the tower.

The inlet water nozzle for back-flushing should be placed below the screen in the bottom of the tower.

The appearance of the glycol can generally be used to determine when the carbon needs to be regenerated or replaced.

The pressure drop across the carbon bed can also be used.

The pressure drop normally across the carbon bed is only 1 or 2 lbs.

When the pressure drop reaches 10 to 15 lbs, the carbon is usually completely plugged with impurities.

Steam cleaning can sometimes be used to regenerate the carbon by removing the impurities.

However, this can be hazardous and offers only limited success.

Another method of purification is to use activated carbon in elements, such as Peco-Char.

Either purification system should be placed downstream from the solids filter.

This will increase the carbon adsorptive efficiency and life.

REFERENCES

Part 1 Hydrate Prediction and Prevention

Arnold, K., & Stewart, M. (1995). *Surface production operations: Design of gas-handling systems and facilities.* Houston: Gulf Publishing Co. Chapter 4.

Karge, F. (1945). Design of Oil Pipelines. *Petroleum Engineer*, (May).

McKetta, J. J., & Wehe, A. H. (1958). Use this chart for water content on natural gases. *Petroleum Refiner*, (August), 153.

Minkkinen, A., et al. (1992). Methanol gas-treating scheme offers economics, versatility. *Oil and Gas Journal*, (June), 65.

National Tank Company (1958). *Engineering Methods:* Specific Heat Factors-Temperature vs. Pressure, Tulsa OK.

Nielsen, R. B., & Bucklin, R. W. (1983). Why not use methanol for hydrate control. *Hydrocarbon Processing*, (April), 71.

Part 2 Dehydration Considerations

Arnold, K., & Stewart, M. (1991). *Surface production operations: Design of oil-handling systems and facilities.* Houston: Gulf Publishing Co. Chapters 4 and 5.

Arnold, K., & Stewart, M. (1995). *Surface production operations: Design of gas-handling systems and facilities.* Houston: Gulf Publishing Co. Chapter 8.

Ballard, D. (1966). How to operate a glycol plant. *Hydrocarbon Processing*, (June), 180.

Dehydration. In *Engineering data book* (11th ed.). Sec. 19, 20, & 21. (1998). Gas Processor's Suppliers Association/Gas Processors Association, Tulsa.

Gas Conditioning Fact Book. (1962). Dow Chemical of Canada Ltd., Toronto.

Grosso, S. (1978). Glycol choice for gas dehydration merits close study. *Oil and Gas Journal*, (February), 106.

Holder, M.R (1991). Performance troubleshooting on a TEG dehydration unit with structured packing. In *Proc., Laurence Reed Gas Conditioning Conference* (pp. 100–112). Oklahoma: Norman.

Kean, J. A., Turner, H. M., & Price, B. C. (1991). How packing works in dehydrators. *Hydrocarbon Processing*, (April), 47.

Kraychy, P. N., & Masuda, A. (1966). Molecular sieves dehydrate high-acid gas at pine creek. *Oil and Gas Journal*, (August), 66.

McKetta, J. J., & Wehe, A. H. (1958). Use this chart for water content on natural gases. *Petroleum Refiner*, (August), 153.

Redus, F. R. (1966). Field operating experience with calcium chloride gas dehydrators. *World Oil*, (February), 63.

Personal communication and trouble-shooting with numerous field operations personnel with Total E&P Indonesie, Chevron Nigeria Limited, Unocal Indonesia, Unocal Thailand, PTTEP, Cabinda Gulf Oil Company, Petronas, Brunei Shell, Sarayak Shell, BP Indonesia, BP Viet Nam, and ExxonMobil Malaysia.

Part III Glycol Maintenance, Care, and Troubleshooting

Personnel communication with Rocky Buras, President; Mark Middleton, Vice President and other senior Gly-Tech personnel, 2054 Paxton Street, Harvey, LA. 70058.

INDEX

Note: Page numbers followed by *f* indicate figures and *t* indicate tables.

Printed and bound by CPI Group (UK) Ltd, Croydon, CR0 4YY

03/10/2024

01040435-0002